Statistical Quality Assurance

Statistical Quality Assurance

by
Francis J. Guldner

DELMAR PUBLISHERS INC.®

Administrative editor: Mark W. Huth

For information, address Delmar Publishers Inc.
2 Computer Drive West, Box 15-015
Albany, New York 12212

Printed in the United States of America
Published simultaneously in Canada
by Nelson Canada,
A division of International Thomson Limited

9 8 7 6 5 4 3 2

Library of Congress Cataloging-in-Publication Data

Guldner, Francis J.
Statistical quality assurance.

Includes index.
1. Quality control—Statistical methods. I. Title.
TS156.G85 1986 658.5′62 86-6387
ISBN 0-8273-2665-3
ISBN 0-8273-2666-1 (instructor's guide)

Contents

Preface

Quality assurance is not the process of throwing out the rejects and shipping the satisfactory. The purpose of statistical quality assurance is to prevent the production of unsatisfactory goods. In order to do this, quality assurance techniques must be integrated in the production process.

This text is designed for a course in statistical quality assurance. Such a course was included only in Industrial Engineering curriculum. It is now a standard course in many Engineering and Management curricula, and separate certificate programs.

Prerequisites

It is assumed that the reader has completed at least one semester or quarter of college algebra. Prior knowledge of statistics or probability is not required.

Major Features

The minimum level of statistics and probability required to apply statistical quality assurance techniques are included in the text. Non-essential nomenclature and concepts are not included.

There are at least two complete examples provided for each type of calculation as well as ample practice problems at the end of each chapter. Step by step procedures are provided for the application of each technique.

The summary of each chapter includes an explanation of unique terms and a list of formulas used in the chapter.

The major part of the L.L. Bean inspection plan is provided as an example of an acceptance inspection plan.

Use is made of standard production planning documentation to integrate quality assurance in the production process.

BASIC language programs which do statistical quality assurance calculations are included in an appendix.

Organization of the Text

The first portion of the text covers the statisical techniques required to determine the process capability. It proceeds to the use of sample data to estimate the process capability, and the development and use of control charts.

The next step is the understanding of simple probability. This leads to the determination of sample size and frequency for useful control charts.

Control charts for attributes are then developed. This is followed by acceptance sampling and the use of standard acceptance sampling plans.

The final chapter covers the integration of quality assurance in the production process.

Acknowledgments

This book contains the minimum of statistical nomenclature and mathematics necessary due to my daughter, Frances D. Guldner, who zealously attacked any unusual word or calculation and insisted that the sentences be understandable.

For technical accuracy and complete coverage of the subject, I am indebted to five people who provided most thorough technical reviews of the first draft. They are:

Ernest Balfrey, Balfrey Associates
William H. Brennan, Rexnord Inc.
Ted Flemming, Tarrant County Junior College
Robert Rosenfeldt, Ford Motor Company
Susan Stewart, Clackamas Community College

For continued support and guidance on getting the ideas on paper, I thank Mark Huth of Delmar Publishers.

Introduction

Quality. As defined in the *American Heritage Dictionary* is "a characteristic or attribute of something."

Statistics. As defined in the *American Heritage Dictionary* is "the mathematics of the collection, organization, and interpretation of numerical data."

Control. As defined in the *American Heritage Dictionary* is "the ability to regulate."

Statistical Quality Assurance. Our working definition will be "the use of the mathematics of the collection, organization, and the interpretation of numerical data to regulate a characteristic or attribute of something." The word *regulate* means that the management of the process is an essential part of quality control. This requires close coordination and interaction among those manipulating the data and those manipulating the process.

Characteristic. We will use the term *characteristic* for measurable quantities such as length, volume, weight, and temperature. These quantities can have a large number of possible measurements. The diameter of shafts produced from a single machine may vary from 1.00001 inches to 1.01 inches, one thousand possible measurements of 0.00001-inch accuracy.

Attribute. We will use the term *attribute* for things that are measured in integers or counted (1, 2, 3, etc.). The number of missing rivets in an aircraft wing may be one or six of any whole number but cannot be 1.25. The attribute of rivet is either present or not.

Process Capability. We consider a process to be the combination of material, machines, and operators that produces a characteristic or attribute. A bar of steel, a lathe, and an operator may produce a shaft with the characteristic of a 1-inch diameter. Due to normal variations in the process, such as looseness in the lathe, all diameters will not be 1.000000 inch. Some will be slightly larger and some slightly smaller. This variation in the characteristic diameter, which occurs when everything is working normally, is called the *process capability*.

Specifications. Most product designs do not require that a stated dimension (characteristic) be maintained exactly accurately within the ability to measure. The allowable variation from the stated dimension is called the *tolerance* and is given for larger (+) and smaller (−) measurements. The size of the tolerance is determined by the use of the part. The diameter of a shaft to be used as the axle for a lawn mower wheel may have the specification 0.50 ± 0.05 inch (0.50 plus or minus 5 hundredths inch). A similar shaft to be used in a fuel injector motor may have a specification of 0.50 ± 0.00005 inch. The manufacturing process is determined by the specifications. The lawn mower axle could be made on a lathe. The injector shaft could not be made on a lathe because the process capability of a lathe has more than 0.00005 inch variation; the diameter could not be held within the specifications.

The Relationship of Statistics, Process Capability, and Specifications. We can measure the characteristic (e.g., diameter) of parts produced by a given process. We can then use statitstics to interpret this measurement data and predict the percentage of parts that will be within any size range. For example, we could say that 95% of all parts will be within 0.995 inch and 1.005 inches and that 10% will be larger than 0.980 inch. We can use statistics to fully describe the process capability. This information on the process capability can be used to determine if a given process will produce parts that meet given specifications. If all parts will not meet the specifications, we can determine the number that would be too small or too big. If the characteristic is the outside diameter of a shaft, those that are too big can be reworked (machined again); those too small are scrap. We can use our knowlege of the process capability to determine the machine settings that would cause almost all the shafts that do not meet the specifications to be too big, thereby reducing scrap losses. By knowing the approximate number of shafts requiring rework, this work can be scheduled in advance.

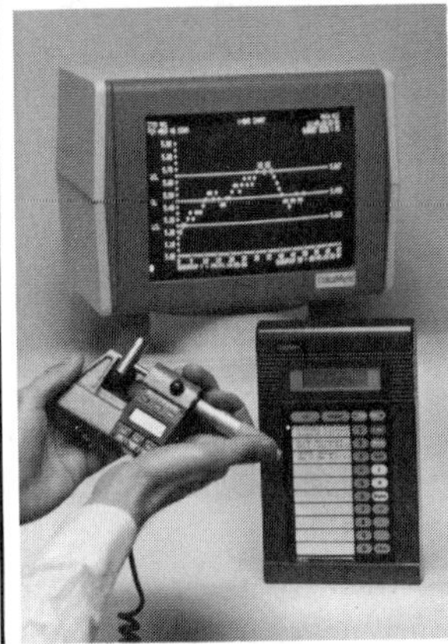

The DataMyte quality assurance system is called FAN, Factory Area Network. It provides a combination of immediate data analysis to the operator and batch data collection for management action. The heart of the system is the DataMyte 750.

The DataMyte 750 is a rugged data collector and statistical computer. It can store up to 2500 readings from micrometers, calipers, dial indicators, and many other gauges. Statistical computations are displayed on it by pressing a single key. Readings are compared against specifications or control limits with instant feedback. Date and time of the sample readings are automatically recorded. The 750 outputs a complete $\bar{X}$ and R chart or histogram when connected to a video monitor. All this information is available to the machine operator without a single calculation, without transcription errors, and without any paperwork.

The data from the DataMyte 750s can be transmitted to a DataTruck, which is similar in appearance to the 750 but has up to 128k of memory. This DataTruck can produce other printed reports and communicate with a computer that is capable of providing various reports including scrap estimates and Pareto analysis.

The unique characteristic of the DataMyte FAN system is the emphasis of data collection and instant feedback at the point of production. This requires some additional training of production workers, but it provides the production worker with the information needed to maintain quality. The worker can take additional samples and get statistical analysis when he or she thinks the process is changing rather than waiting for others to do it on schedule or on request.

The Relationship of Process Capability and Quality Control. Once we know the process capability we can monitor the production and determine when the process has changed. If every hour we take five pieces from the production line and measure the diameter of each, we can calculate the average diameter of the sample of five. Our knowledge of the process capability will let us calculate the likely minimum and maximum values that a sample average would have if the process continued unchanged. When the sample average is outside these values, we have reason to believe that the process has changed and needs correction. This change in the process may be a change in material, a change in the operator's procedure, or a change in the machine (e.g., tool wear). This statistical sampling of production will allow us to take corrective action before producing rejects. It will also prevent us from wasting time correcting the process when it is performing correctly.

Organization of the Text. The first portion covers the statistical techniques needed to determine the process capability by measuring every part produced. Then we'll see how to estimate the process capability by measuring only some of the parts produced. Learning how to calculate the maximum and miminum values to be expected for a sample average will let us determine the procedures to alert us that the process has changed.

The next step is to learn the laws of probability needed to do the same interpretation of data on attributes that we did on characteristics.

The final statistical portion of the text covers taking samples of items received from suppliers to see if the items received meet our specifications.

In addition to counting rejected items, we can keep track of the reasons for rejection. This allows for the discovery of the primary cause of a production problem in a logical manner.

Statistical Quality Assurance

1 Basic Structure

This chapter covers the organization of numerical data into graphic format. This picture of the measurement data reveals the process capability. We will use the tool of statistics (the mathematics of the collection, organization, and intepretation of numerical data) to tell us what the process is producing. With this information it is possible to make decisions about modifying the process to improve quality and reduce costs. Knowing the extreme values produced when the process is working normally allows early identification of process changes by looking for values outside these normal extremes.

Objectives

After completing this chapter you will be able to:

- organize measurement data into a frequency chart.
- calculate a data mean and standard deviation.
- determine process capabilities.
- compare process capability with specifications.
- calculate a process mean shift to improve compliance with specifications.

Statistics

Statistics is the collection, organization, and interpretation of numerical data.

Let's say that we are manufacturing thermal switches that should trip at 12°F. Once the process is running to our satisfaction, we measure the actual trip temperature of 100 switches. The measurements are rounded to the nearest whole number. Measurements between 11.5 and 12.5 are counted as 12°F. These measurements are shown in Table 1.1.

To organize the data so that it gives a better picture of the process, we will count the number of times each measurement occurred. The value 6°F appears once. There were no 7°F measurements. 8°F appears twice, and so on. The measurement values are listed in numerical order and the number of times that measurement appears, the *frequency* is listed next to it. This gives us a *frequency table*, Table 1.2.

Table 1.1

Trip Temperatures in °F.					
13	16	16	13	13	9
12	15	8	12	13	14
11	6	15	11	15	13
12	12	13	13	13	11
17	14	11	14	13	15
11	11	13	15	9	10
11	12	11	16	11	10
13	13	11	13	10	13
11	10	8	11	11	17
10	12	13	10	10	15
14	11	9	15	13	12
11	14	12	10	12	14
14	12	11	11	12	13
11	11	11	13	12	11
12	10	9	12	10	13
12	14	10	10	16	12
11	11	14	14		

The final step is to add a bar chart or histogram to the frequency chart. As shown in Table 1.3, the length of the bar graphically displays the frequency value.

When the data is organized and displayed in this manner, it is easy to see that the trip temperature of our switches goes from 6°F to 17°F with the bulk of them in the 10°F to 14°F range.

This distribution of measurements is the *process capability* as defined in the introduction. Although the average trip temperature of our switches is 12°F, very few of the switches trip at exactly 12°F. This variation of measurements that takes place when the manufacturing process is working normally is what our process is capable of producing. This process capability can be compared with specification limits to determine if the process is capable of meeting design specifications.

By adding the specification limits to the chart, we can quickly determine whether or not we have a problem. The specification limits of 12°F ± 2°F are added to the frequency distribution in Table 1.4.

By adding the frequencies from 10°F to 14°F, we see that only 81% of our production would meet the specification 12°F ± 2°F, but 100% would meet the specification 12°F ± 7°F.

Averages and Measure of Dispersion

Although the frequency chart gives a good picture of the distribution of measurement data, it is time consuming to prepare. There are two mathematical measures that will often give the required information about the data. The *mean* (average) of the values will provide the center of the data. The mean alone is not a sufficient description of the data. The mean of 1 and 11 is 6; the mean of 5 and 7 is also 6. To fully describe the data, we require a measure of the dispersion of the data. In quality control the most common measure of dispersion is the *standard deviation*.

Table 1.2 Frequency Table

°F	*Frequency*	*°F*	*Frequency*
6	1	12	17
7	0	13	19
8	2	14	10
9	4	15	7
10	12	16	3
11	23	17	2

Table 1.3 Frequency Table with Bar

°F	*Frequency*	*Bar*
6	1	I
7	0	
8	2	II
9	4	IIII
10	12	IIIIIIIIIIII
11	23	IIIIIIIIIIIIIIIIIIIIIII
12	17	IIIIIIIIIIIIIIIII
13	19	IIIIIIIIIIIIIIIIIII
14	10	IIIIIIIIII
15	7	IIIIIII
16	3	III
17	2	II

The *mean* is the total of the measurement divided by the number of measurements and is represented as X bar, $\bar{X}$. The mean of the 100 measurements in Table 1.1 is the total of the 100 measurements, 1209, divided by the number of measurements, 100.

$$\bar{X} = \frac{13 + 12 + 11 + 12 + \ldots + 12}{100}$$

$$\bar{X} = 12.09$$

The *standard deviation*, which we will call sigma, σ, is calculated by the formula:

$$\sigma = \sqrt{\Sigma \frac{(X - \bar{X})^2}{n}}$$

In this formula each measured value is **X**. The number of values is **n**. The difference between each value, X, and the mean $\bar{X}$, is squared and the results totaled. This total is then divided by the number of measurements. The square root

Table 1.4 Frequency Table with Specification Limits

°F	*Frequency*	*Bar*
6	1	I
7	0	
8	2	II
9	4	IIII
10	12	IIIIIIIIIIII
11	23	IIIIIIIIIIIIIIIIIIIIIII
12	17	IIIIIIIIIIIIIIIII
13	19	IIIIIIIIIIIIIIIIIII
14	10	IIIIIIIIII
15	7	IIIIIII
16	3	III
17	2	II

of this number is the standard deviation. For the data in Table 1.1, the standard deviation is calculated to be 1.985°F.

$$\sigma = \sqrt{\Sigma \frac{(13-12.09)^2 + (12-12.09)^2 + \ldots + (12-12.09)^2}{100}}$$

$$\sigma = 1.985$$

Many pocket calculators have the calculation of sigma as a built-in function.

The mean and the standard deviation can also be calculated from the frequency chart data (Table 1.5). In the formulas below, **f** is the frequency, **X** is the measurement value, **n** is the number of measurements, and $\bar{X}$ is the mean.

$$\sigma = \sqrt{\Sigma \frac{fX^2}{n} - \bar{X}^2}$$

Table 1.5 Frequency Table for Calculating Mean and Standard Deviation

Value X X	*Frequency f* f	*f times X* fX	*X squared* X^2	*f times X squared* fX^2
6	1	6	36	36
7	0	0	49	0
8	2	16	64	128
9	4	36	81	324
10	12	120	100	1200
11	23	253	121	2783
12	17	204	144	2448
13	19	247	169	3211
14	10	140	196	1960
15	7	105	225	1575
16	3	48	256	768
17	2	34	289	578
TOTALS	100	1209		15011

The total of the frequencies, 100, is the total number of measurements. The total of the **fX** column, 1209, is the total of all the measurements.

$$\bar{X} = \Sigma \frac{fX}{n} = \frac{1209}{100} = 12.09$$

$$\sigma = \sqrt{\Sigma \frac{fX^2}{n} - \bar{X}^2}$$

$$\sigma = \sqrt{\frac{15011}{100} - (12.09)^2}$$

$$\sigma = 1.985$$

Grouping Measurement Data

The measurements in the previous example are whole numbers and there are only 12 values (6 thru 17). If the measurements had been made to one decimal (e.g., 9.1), there would have been 120 possible values (e.g., 8.1°F, 8.2°F) with very few appearing more than three times. The frequency chart was made by grouping measurements into cells. The values between 8.50 and 9.49 were counted in cell

Table 1.6 Fluid in 32-Ounce Containers

Measurements in Ounces					
32.112	32.146	32.097	32.114	32.116	32.062
32.1	32.141	32.056	32.102	32.117	32.122
32.092	32.03	32.142	32.083	32.132	32.113
32.099	32.103	32.11	32.115	32.107	32.085
32.16	32.127	32.089	32.131	32.117	32.139
32.091	32.094	32.114	32.141	32.06	32.076
32.085	32.098	32.092	32.156	32.089	32.076
32.109	32.111	32.085	32.115	32.081	32.118
32.083	32.079	32.056	32.082	32.091	32.161
32.08	32.103	32.118	32.073	32.075	32.136
32.123	32.083	32.062	32.132	32.115	32.104
32.087	32.12	32.104	32.078	32.097	32.127
32.122	32.094	32.088	32.081	32.105	32.118
32.083	32.085	32.085	32.115	32.102	32.091
32.098	32.075	32.066	32.103	32.076	32.114
32.099	32.122	32.08	32.071	32.145	32.102
32.093	32.087	32.126	32.121		

9. The numbers 8.50 and 9.49 are the *cell boundaries*. The number 9 is the *cell midpoint*. The effect is that of rounding off the values to the nearest integer, which is how the measurements in Table 1.1 are shown. The cell boundaries have one more decimal place than the cell midpoints. This insures that every measurement falls into a cell and not on a boundary.

Table 1.6 contains one hundred measurements of the fluid in quart containers. The measurements were taken to one thousandth of an ounce.

Table 1.7 lists these values rounded off to two decimal places. The cell midpoints have two decimal places; for example, 32.03. The cell boundaries have three decimal places. The cell boundaries for the cell that has 32.03 as a midpoint are 32.025 and 32.034. Any measurements from 32.025 thru 32.034 are rounded to 32.03

Table 1.7 Fluid Frequency Table with Bar

Cell Midpoint	*Frequency*	*Graph*
32.02	0	
32.03	1	I
32.04	0	
32.05	0	
32.06	5	IIIII
32.07	3	III
32.08	16	IIIIIIIIIIIIIIII
32.09	18	IIIIIIIIIIIIIIIIII
32.1	15	IIIIIIIIIIIIIII
32.11	10	IIIIIIIIII
32.12	16	IIIIIIIIIIIIIIII
32.13	6	IIIIII
32.14	5	IIIII
32.15	2	II
32.16	3	III

Table 1.8 lists the frequency data in Table 1.7 and shows the values calculated for fX, X^2, and fX^2. The totals of f, fX and fX^2 are used to calculate the mean and standard deviation.

$$\bar{X} = \Sigma \frac{fX}{n} = \frac{3210.22}{100} = 32.102$$

$$\sigma = \sqrt{\Sigma \frac{fX^2}{n} - \bar{X}^2} = \sqrt{\frac{103055.18}{100} - 1030.5384}$$

$$\sigma = 0.02356$$

Table 1.8 Frequency Table for Calculating Mean and Standard Deviation

Value X *X*	*Frequency f* *f*	*f times X* *fX*	*X squared* X^2	*f times X squared* fX^2
32.02	0	0	1025.2804	0
32.03	1	32.03	1025.9209	1025.9209
32.04	0	0	1026.5616	0
32.05	0	0	1027.2025	0
32.06	5	160.3	1027.8436	5139.218
32.07	3	96.21	1028.4849	3085.4547
32.08	16	513.28	1029.1264	16466.0224
32.09	18	577.62	1029.7681	18535.8258
32.1	15	481.5	1030.41	15456.15
32.11	10	321.1	1031.0521	10310.521
32.12	16	513.92	1031.6944	16507.1104
32.13	6	192.78	1032.3369	6194.0214
32.14	5	160.7	1032.9796	5164.898
32.15	2	64.3	1033.6225	2067.245
32.16	3	96.48	1034.2656	3102.7968
TOTALS	100	3210.22		103055.1844

The procedure for developing a frequency chart is as follows:

- Select the number of decimal places to be included in the chart.
- Round off the measurement values to the selected number of decimal places.
- List the different measurement values in numerical order.
- Count the number of times each measurement appears and put that frequency next to the measurement.
- If a graph is desired, draw a bar to scale.

Table 1.9 Fluid in 32-Ounce Containers

32.107	32.128	32.098	32.108	32.110	32.077
32.1	32.125	32.074	32.101	32.110	32.113
32.095	32.058	32.125	32.090	32.119	32.108
32.099	32.102	32.106	32.109	32.104	32.091
32.136	32.116	32.093	32.119	32.110	32.123
32.095	32.096	32.108	32.124	32.076	32.086
32.091	32.099	32.095	32.133	32.093	32.085
32.105	32.106	32.091	32.109	32.089	32.111
32.09	32.088	32.074	32.089	32.095	32.136
32.088	32.102	32.111	32.084	32.085	32.121
32.114	32.09	32.077	32.119	32.109	32.102
32.092	32.112	32.102	32.087	32.098	32.116
32.113	32.096	32.093	32.089	32.103	32.111
32.09	32.091	32.091	32.109	32.101	32.095
32.099	32.085	32.08	32.102	32.086	32.108
32.099	32.113	32.088	32.082	32.127	32.101
32.096	32.092	32.116	32.112		

The measurements listed in Table 1.9 are also of the fluid in 32-ounce containers. Table 1.10 shows these values rounded to two decimal places and listed in numerical order of the cell midpoints. The column **Frequency** shows the number of times each rounded value appears.

The procedure to calculate the standard deviation from frequency chart data is as follows. (The results of these calculations are shown in Table 1.11.)

- List the two data columns of the frequency chart: the cell midpoint value **X**, and the frequency, **f**.
- Add column **fX**, the cell midpoint times the frequency.
- Add column $\mathbf{X^2}$, the cell midpoint squared.

Table 1.10 Fluid Frequency Table with Bar

Cell Midpoint	*Frequency*	*Bar*
32.06	1	I
32.07	2	II
32.08	6	IIIIII
32.09	26	IIIIIIIIIIIIIIIIIIIIIIIIII
32.10	25	IIIIIIIIIIIIIIIIIIIIIIIII
32.11	24	IIIIIIIIIIIIIIIIIIIIIIII
32.12	9	IIIIIIIII
32.13	5	IIIII
32.14	2	II

- Add column fX^2, the frequency times the squared cell mid-point.
- Total columns **f, fX**, and fX^2.
- Calculate the mean, $\bar{X}$, using the formula

$$\bar{X} = \Sigma \frac{fX}{n}$$

- Calculate the standard deviation, σ, using the formula

$$\sigma = \sqrt{\Sigma \frac{fX^2}{n} - \bar{X}^2}$$

$$\bar{X} = \Sigma \frac{fX}{n} = \frac{3210.17}{100} = 32.1017 \; oz.$$

$$\sigma = \sqrt{\Sigma \frac{fX^2}{n} - \bar{X}^2}$$

$$\sigma = \sqrt{\frac{103051.9361}{100} - (32.1017)^2}$$

$$\sigma = 0.01604 \; oz.$$

Table 1.11 Calculation of Standard Deviation

Value X *X*	*Frequency f* *f*	*f times X* *fX*	*X squared* *X²*	*f times X squared* *fX²*
32.06	1	32.06	1027.8436	1027.8436
32.07	2	64.14	1028.4849	2056.9698
32.08	6	192.48	1029.1264	6174.7584
32.09	26	834.34	1029.7681	26773.9706
32.1	25	802.5	1030.4100	25760.2500
32.11	24	770.64	1031.0521	24745.2504
32.12	9	289.08	1031.6944	9285.2496
32.13	5	160.65	1032.3369	5161.6845
32.14	2	64.28	1032.9796	2065.9592
TOTALS	100	3210.17	9273.6960	103051.9361

Statistical Quality Assurance

Statistical quality assurance is the use of the mathematics of the collection, organization, and the interpretation of numerical data to regulate the characteristic or attribute of something. The interpretation of the data has described the process capability. Regulating the process consists of two activities:

1. Modifying the process.
2. Monitoring the process to detect changes.

Modifying the process may be as extreme as a complete change such as determining that a lathe process capability will not meet the requirement and the job must be done on a grinding machine to meet specifications. It may also be as simple as an adjustment of machine settings to change the mean value of a characteristic.

Monitoring the process consists of measuring the output to detect changes. Since we now have determined the extreme values of the process when it is operating normally, any value outside of this range would indicate a possible change in the process and be justification for a possible process change. With the same information, we know which extreme values will occasionally occur, and we will not stop the process to make changes while it is still functioning normally.

Interpretation of the Data

From the organized data in Table 1.7 and calculations from Table 1.8, we can see that the process is providing 100% satisfaction of a requirement to provide at least 32 ounces in a quart. With the mean of 32.10 ounces, we can see that overfilling of the containers is 0.3 percent. The containers contain from 32.03 ounces to 32.16 ounces.

The cost of this overfill can be calculated and decisions made about the economy of modifying the process to reduce the standard deviation. Examination of the frequency chart shows that a slight reduction of the overfill is possible by shifting the mean to 32.08 or 32.07, if this can be done without changing the standard deviation. If the data had shown some containers underfilled, a mean shift to insure 100% fill could be calculated in a similar manner.

Now that it is determined that the process produces containers with a minimum 32.03 ounces and a maximum of 32.16 ounces, future production can be compared with these values. If the process starts producing containers with 32.01 ounces, we have reason to believe that the process has changed, and we investigate because 32.01 ounces is significantly below the normal process minimum. This will allow correction of the process before underfilled or reject containers are produced.

Chapter Review

In this chapter we have seen the use of statistics to determine the process capabilities in two cases where every item produced was measured. This 100% measurement is costly and is not always necessary. It is possible to estimate the process capability and monitor production by taking samples. In the next chapter we will compare the actual data in these two cases with theoretical data based only on the mean and standard deviation. We will then determine how to get a satisfactory picture of the process capability by taking samples of the output instead of measuring all of the production.

Keywords

Frequency chart A table showing the values of a distribution with the tally of the number of times each value occurred. It normally is represented by a bar graph or histogram.

Cell Midpoint The value used in a frequency chart to represent actual measurement values within the cell boundaries.

Cell boundaries The maximum and minimum values to be counted in a single cell.

Specification The upper and lower limits of a characteristic that will meet the requirements of the user.

Mean The arithmetical average of a set of numbers.

Standard deviation A calculated measure of the dispersion of a set of numerical values.

Formulas

Mean $\bar{X}$

From individual values:

$$\bar{X} = \frac{\textit{total of measurements}}{n}$$

From a frequency chart:

$$\bar{X} = \Sigma \frac{fX}{n}$$

Standard deviation σ

From individual values:

$$\sigma = \sqrt{\Sigma \frac{(X - \bar{X})^2}{n}}$$

From a frequency chart:

$$\sigma = \sqrt{\Sigma \frac{fX^2}{n} - \bar{X}^2}$$

Problems

1-1 A company produces one-inch diameter steel shafts for electric motors. After the shaft grinding process was set up, the first 100 shafts were measured. (See Table 1.12).

a. Make a frequency chart and bar chart of these measurements. Group the data by rounding to two decimal places.

b. Calculate the mean of the data.

c. Calculate the standard deviation of the data.

1-2 If the specifications required that all shafts in problem 1-1 have a diameter from 1.00 to 1.12 inches (1.06 ± 0.06 in.), what percentage of production would be out of specification? If the mean could be shifted without changing the standard deviation, what new mean would you recommend? What percentage of production would be out of specification? Would they be too big or too small?

Table 1.12 Steel Shaft Diameter: First Machine

1.061	1.111	1.116	1.085	1.097	1.1
1.083	1.052	1.103	1.118	1.09	1.072
1.1	1.092	1.07	1.133	1.134	1.075
1.107	1.102	1.127	1.121	1.086	1.093
1.125	1.105	1.115	1.097	1.15	1.107
1.079	1.104	1.049	1.129	1.111	1.104
1.142	1.099	1.095	1.088	1.076	1.095
1.091	1.07	1.098	1.118	1.132	1.142
1.129	1.073	1.102	1.093	1.141	1.136
1.111	1.063	1.13	1.136	1.116	1.118
1.092	1.074	1.104	1.147	1.125	1.08
1.106	1.067	1.074	1.08	1.083	1.067
1.111	1.128	1.142	1.063	1.108	1.126
1.144	1.12	1.136	1.127	1.047	1.128
1.092	1.133	1.111	1.102	1.097	1.147
1.068	1.108	1.097	1.073	1.079	1.085
1.139	1.137	1.077	1.074		

1-3 A second grinding machine was added to meet increased production requirements. The diameters of the first 100 shafts are shown in Table 1.13.

a. Make a frequency chart and bar chart of these measurements. Group the data by rounding to two decimal places.

b. Calculate the mean of the data.

c. Calculate the standard deviation of the data.

1-4 If the specifications required that all shafts in problem 1-3 have a diameter from 1.00 to 1.12 inches (1.06 ± 0.06 in.), what percentage of production would be out of specification? If the mean could be shifted without changing the standard deviation, what new mean would you recommend? What percentage of production would be out of specification? Would they be too big or too small?

Table 1.13 Steel Shaft Diameter: Second Machine

1.111	1.112	1.103	1.104	1.111	1.115
1.116	1.071	1.106	1.109	1.075	1.096
1.053	1.109	1.125	1.105	1.111	1.102
1.086	1.1	1.06	1.129	1.102	1.08
1.135	1.126	1.068	1.098	1.092	1.108
1.155	1.129	1.088	1.139	1.107	1.075
1.107	1.107	1.069	1.088	1.105	1.101
1.111	1.167	1.115	1.085	1.103	1.089
1.07	1.072	1.116	1.131	1.101	1.101
1.093	1.119	1.091	1.11	1.105	1.079
1.074	1.089	1.146	1.136	1.13	1.1
1.113	1.118	1.092	1.133	1.149	1.1
1.094	1.077	1.099	1.133	1.117	1.119
1.059	1.094	1.118	1.099	1.073	1.059
1.081	1.131	1.086	1.129	1.062	1.102
1.093	1.105	1.12	1.102	1.09	1.089
1.086	1.119	1.096	1.081		

2 Normal Distribution

This chapter covers the normal distribution. It compares the distribution of values in a normal distribution with those in Chapter 1 and the use of the normal distribution to estimate a process capability.

Objectives

After completing this chapter you will be able to:

- describe the normal distribution.
- calculate the area under the normal curve between limits.
- using the values of $\bar{X}$ and σ, determine a process capability.
- using a process $\bar{X}$ and σ, determine the percentage of production between any two limits.
- using a process $\bar{X}$ and σ, determine the percentage of production below or above specification limits.
- calculate a process $\bar{X}$ that would maximize the percentage of production within a given specification.

The Normal Distribution

The normal distribution is a mathematically derived curve. It is also called the bell curve, the Gaussian curve, and the probability curve. It is symmetrical with a mean, $\overline{X}$,of zero and a standard deviation, σ, of one. The normal distribution is illustrated in Figure 2.1.

Although the theoretical curve extends from minus infinity to plus infinity, most of the area under the curve lies between minus three standard deviations and plus three standard deviation. The total area under the curve is one. Its mean is $\overline{X}$ and its standard deviation is σ.

The shaded area in Figure 2.2 is the area between -1 and $+1$ standard deviations. This area is 68.26 percent of the total area.

The area between -2 and $+2$ standard deviations is 95.46 percent of the total area as shown in the shaded portion of Figure 2.3.

The area between -3 and $+3$ standard deviations is 99.73 percent of the total area under the curve. This is the shaded portion of Figure 2.4. It is referred to as the area between the limits $\overline{X} \pm 3\,\sigma$.

Interpretation of the Normal Distribution

The area under the normal curve is one. Since 99.73 percent of this area is within the limits $\overline{X} \pm 3\,\sigma$, this area is 0.9973. Note that the X axis is expressed in terms of the number of standard deviations from the mean. It is possible to calculate the area under the curve between any two limits. Table A (Appendix A) gives the area from $-\infty$ to any chosen point expressed in terms of the standard deviation σ. The shaded area of Figure 2.5 is between the limits $-\infty$ and $-1\,\sigma$. Table A shows this area to be 0.1587.

The area from $-\infty$ to $1\,\sigma$ is the shaded area of Figure 2.6. Table A shows this area to be 0.8413.

Calculation of Area Between Limits Given in σ

The procedure to determine the area under the normal curve between two limits given in terms of the number of standard deviations from the mean is as follows:

- Sketch the normal curve and show the area to be calculated.
- Get from Table A the area from $-\infty$ to the upper limit.
- Get from Table A the area from $-\infty$ to the lower limit.
- Subtract the area to the lower limit from the area to the upper limit.

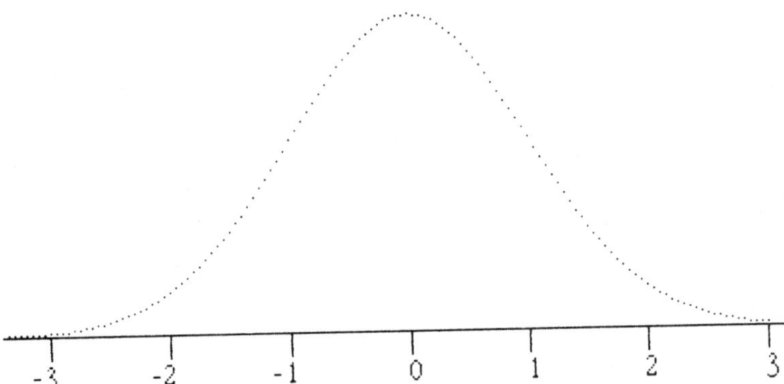

Figure 2.1 Normal Distribution

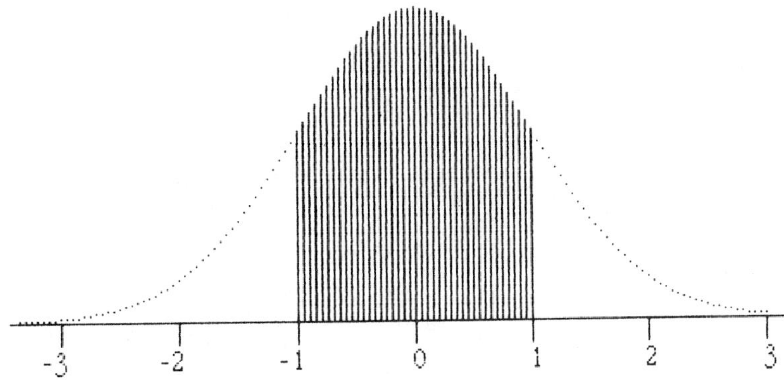

Figure 2.2 Area between -1σ and $+1\sigma$

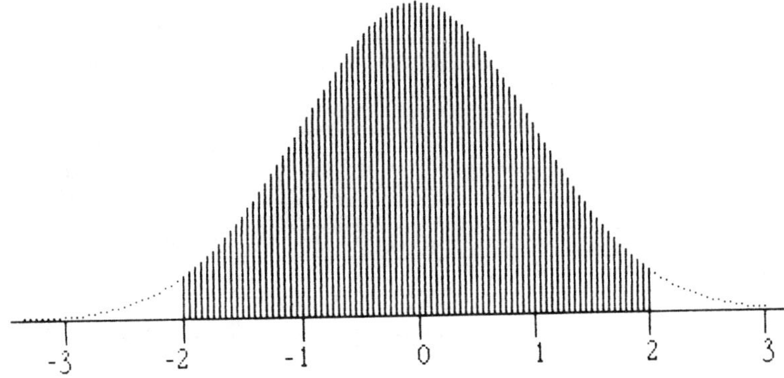

Figure 2.3 Area between -2σ and $+2\sigma$

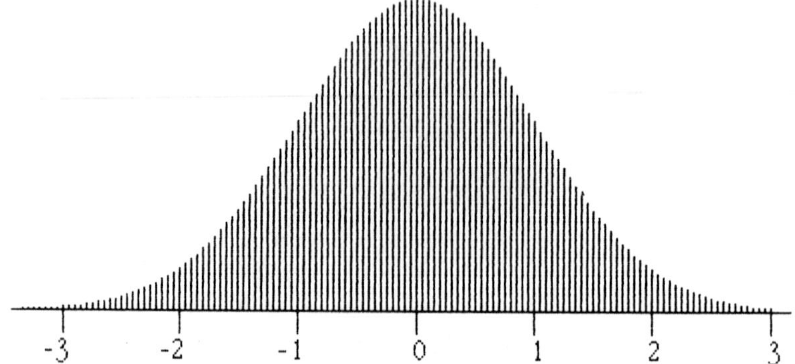

Figure 2.4 Area between the limits $\bar{X} \pm 3\sigma$

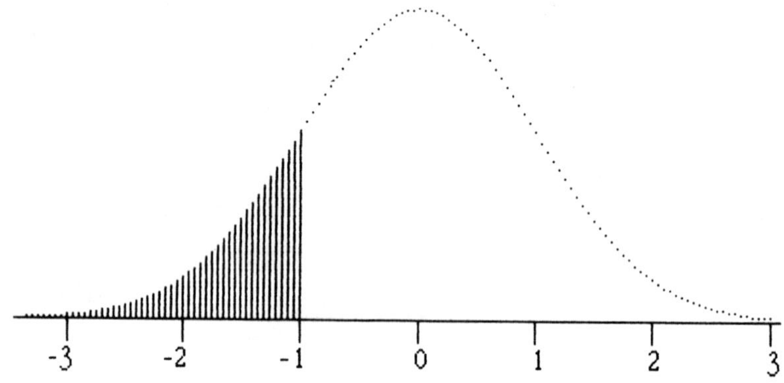

Figure 2.5 Area between the limits $-\infty$ and -1σ

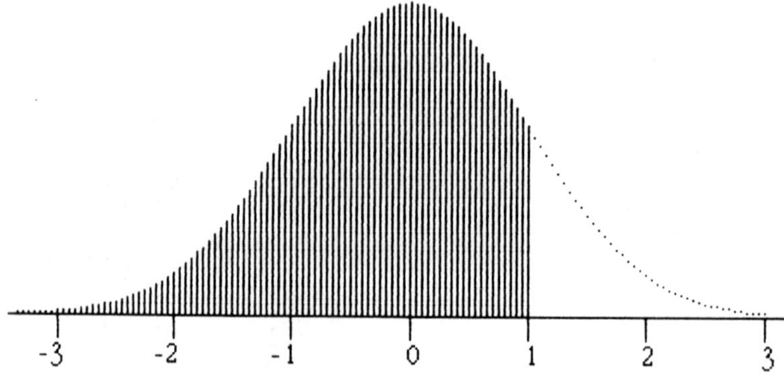

Figure 2.6 Area from $-\infty$ to 1σ

Using this data, Figure 2.5, and Figure 2.6, we can calculate the area between the limits $\bar{X} \pm 1\sigma$.

- Sketch the normal curve and show the area to be calculated (Figure 2.7).
- Get from Table A the area from $-\infty$ to the upper limit.
- Get from Table A the area from $-\infty$ to the lower limit.
- Subtract the area to the lower limit from the area to the upper limit.

The area from $-\infty$ to $+1\sigma$ from Table A = 0.8413.
The area from $-\infty$ to $-\sigma$ from Table A = 0.1587.
The area between the limits $\bar{X} \pm 1\sigma$ = 0.6826.

Example 2-1

Given:

A normal distribution with a mean of 0 and a standard deviation of 1.

Find:

The area under the curve between the limits $\bar{X} - 1.5\sigma$ and $\bar{X} + 2.5\sigma$.

Solution:

Figure 2.8.

The shaded area in Figure 2.8 is between the limits $\bar{X} - 1.5\sigma$ and $\bar{X} + 2.5\sigma$.

The area from $-\infty$ to $+2.5\sigma$ from Table A = 0.9938
The area from $-\infty$ to -1.5σ from Table A = 0.0668
The area between the limits = 0.9270.

Answer:

The area under the normal curve between the limits $\bar{X} + 2.5\sigma$ is 0.9270

The Normal Distribution to Determine Process Capability

Now that we can calculate the area under the normal distribution between any two points, we can apply this method to actual process distributions if we assume that they are almost normally distributed.

In Chapter 1 we determined the mean and standard deviation for the process distributions. In this chapter we determined that 95.46 percent of the area under the normal curve lies between -2 and $+2$ standard deviations. If a process distribution is normally distributed, 95.46 percent of its measurements lie between $\bar{X} - 2\sigma$ and $\bar{X} + 2\sigma$.

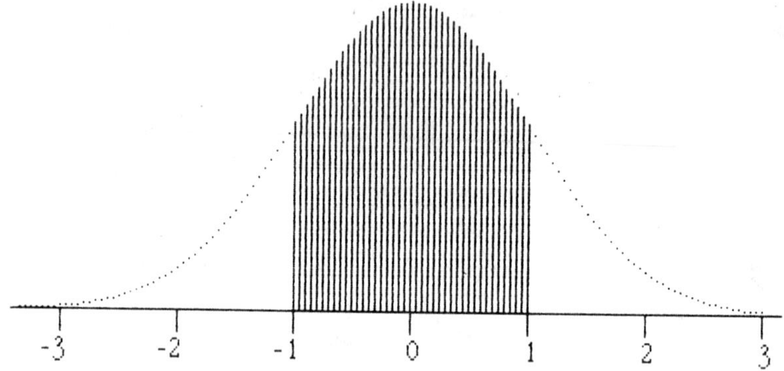

Figure 2.7 Area to be calculated

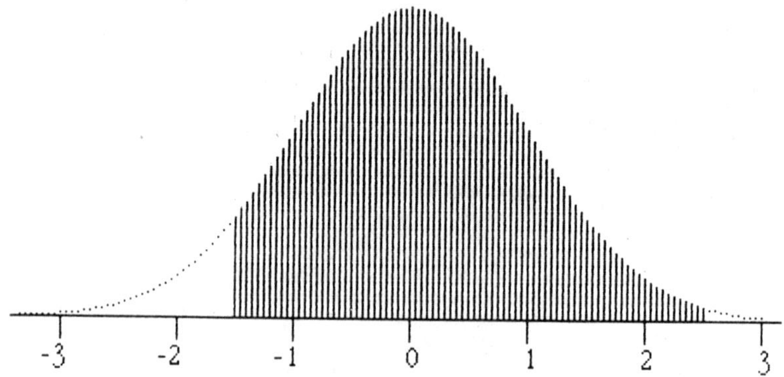

Figure 2.8 Area between the limits $\bar{X} - 1.5\sigma$ and $\bar{X} + 2.5\sigma$.

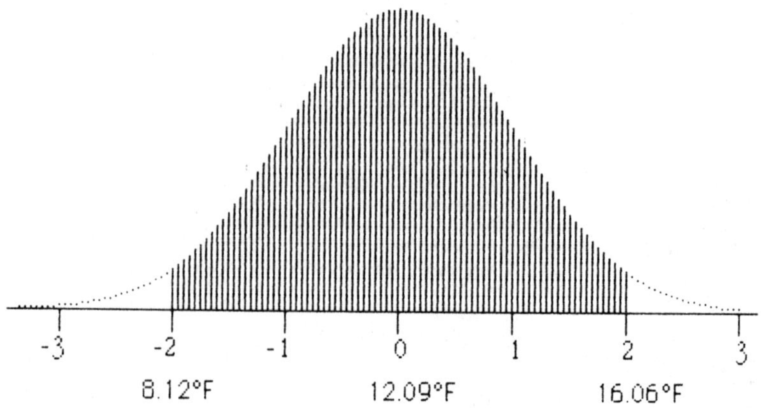

Figure 2.9 Area between $\bar{X} - 2\sigma$ and $\bar{X} + 2\sigma$

From the 100 measurements in Table 1.1 (Chapter 1), we calculated the mean to be 12.09 and the standard deviation to be 1.985. In order to translate the limits expressed as $\overline{X} \pm 2\sigma$ to measurement values, we substitute the values of $\overline{X}$ and σ in the expression $\overline{X} \pm 2\sigma$ and calculate its values. Two standard deviations is two times 1.985, or 3.97. The lower limit, $\overline{X} - 2\sigma$, then becomes 12.09 − 3.97, or 8.12. The upper limit, $\overline{X} + 2\sigma$, becomes 12.09 + 3.97, or 16.06. This is shown in Figure 2.9.

Figure 2.9 shows the normal distribution with a mean of zero and a standard deviation of one. The numbers below these show what our actual process distribution with a mean of 12.09 and a standard distribution of 1.985 would be if its distribution were normal. The lower limit of the shaded area, $\overline{X} - 2\sigma$, has a value of 8.12°F. The upper limit of shaded area, $\overline{X} + 2\sigma$, has a value of 16.06°F.

Table 2.1 shows the actual distribution of this data. The dashed lines indicate the $\overline{X} - 2\sigma$ and $\overline{X} + 2\sigma$ limits. Although the actual distribution is only approximately normal in shape, 95 of the 100 measurements are within the $\overline{X} - 2\sigma$ (8.12°F) and $\overline{X} + 2\sigma$ (16.06°) limits. This is very close to the 95.46 percent calculated from the area under the normal curve.

Table 2.1 Actual Distribution with $\pm 2\sigma$ limits

	°F	*Frequency*	*Bar*
	6	1	I
	7	0	
	8	2	II
8.12	------	------	------
	9	4	IIII
	10	12	IIIIIIIIIIII
	11	23	IIIIIIIIIIIIIIIIIIIIIII
	12	17	IIIIIIIIIIIIIIIII
	13	19	IIIIIIIIIIIIIIIIIII
	14	10	IIIIIIIIII
	15	7	IIIIIII
	16	3	III
16.06	------	------	------
	17	2	II

From Figure 2.8 we calculated that the area under the normal curve between the limits $\bar{X} - 1.5\sigma$ and $\bar{X} + 2.5\sigma$ is 0.9270. If our actual distribution is approximately normal, 92.7 percent of the measurements should lie between those limits. The measurements of these limits are:

$$\bar{X} - 1.5\sigma = 12.09 - 1.5 \times 1.985 = 9.1125°F$$

$$\bar{X} + 2.5\sigma = 12.09 + 2.5 \times 1.985 = 17.0525°F$$

The shaded area of Figure 2.10 shows what these measurements would be if the actual distribution were normal. Table 2.2 shows these limits on the actual distribution. 93 percent of the actual measurements lie between these limits.

Calculation of Area Between Limits Given in Measurement Values

We have demonstrated that we can use the mean and standard deviation of the actual distribution together with the characteristics of the normal distribution to estimate the details of the actual distribution. In Table 1.4 of Chapter 1 we counted the actual number of thermal switches that met the specification of 12°F ± 2°F. To do this we needed to know the actual values of each of the 100 switches.

The number of switches that meet the specification 12°F ± 2°F can be estimated by the use of the mean and standard deviation of the actual distribution and the characteristics of the normal distribution.

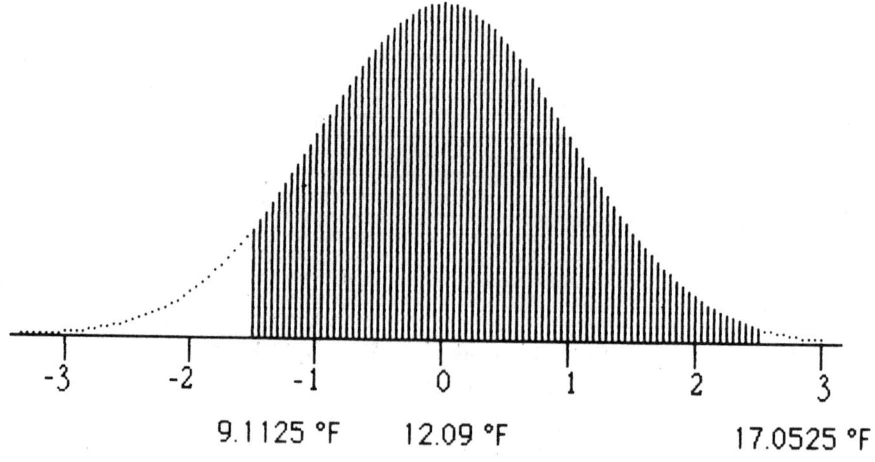

Figure 2.10 Area between $\bar{X} - 1.5\sigma$ and $\bar{X} + 2.5\sigma$

The procedure to determine the area under the normal curve between two limits given in measurement values is as follows:

- Sketch the normal curve and show the area to be calculated.
- Calculate the number of standard deviations between the mean and the upper limit.
- Calculate the number of standard deviations between the mean and the lower limit.
- Get from Table A the area from $-\infty$ to the upper limit.
- Get from Table A the area from $-\infty$ to the lower limit.
- Subtract the area to the lower limit from the area to the upper limit.

Table 2.2 Actual Distribution with -1.5σ and $+2.5\sigma$ limits

	°*F*	*Frequency*	*Bar*
	6	1	I
	7	0	
	8	2	II
	9	4	IIII
9.1125 -------			
	10	12	IIIIIIIIIIII
	11	23	IIIIIIIIIIIIIIIIIIIIIII
	12	17	IIIIIIIIIIIIIIIII
	13	19	IIIIIIIIIIIIIIIIIII
	14	10	IIIIIIIIII
	15	7	IIIIIII
	16	3	III
	17	2	II
17.0525 -------			

Example 2-2

Given:

$\bar{X} = 12°F, \sigma = 1.985°F$

Find:

Percentage of measurements meeting the specification 12°F ± 2°F

Solution:

Since the temperature values are rounded to the nearest integer, values between 9.5°F and 14.4°F meet these specifications.

► Sketch the normal curve and show the area to be calculated.

To calculate the area under the normal distribution between two limits, we must know the number of standard deviations between each limit and the mean. We will use Z for this value. The limits will be called X_i. Then Z for each limit can be calculated by:

$$Z = \frac{Xi - \bar{X}}{\sigma}$$

► Calculate the number of standard deviations between the mean and the upper limit.

Given a mean of 12.09 and a standard deviation of 1.985, the number of standard deviations between the mean and the limit 12°F + 2°F, or 14.4°F, can be calculated as follows:

$$Z = \frac{Xi - \bar{X}}{\sigma} = \frac{14.4 - 12.09}{1.985} = 1.16$$

► Calculate the number of standard deviations between the mean and the lower limit.

The number of standard deviations between the mean and the limit 12°F − 2°F, or 9.5°F, can be calculated as follows:

$$Z = \frac{Xi - \bar{X}}{\sigma} = \frac{9.5 - 12.09}{1.985} = -1.3$$

This area is shown as the shaded area in Figure 2.11.

The procedure to determine the area between two limits is:

- Get from Table A the area from $-\infty$ to the upper limit.
- Get from Table A the area from $-\infty$ to the lower limit.
- Subtract the area to the lower limit from the area to the upper limit.

Using this data, we can calculate the area between the limits $\bar{X} + 1.16\sigma$ and $\bar{X} - 1.3\sigma$.

The area from $-\infty$ to $+1.16\sigma$ from Table A $= 0.8770$
The area from $-\infty$ to -1.3σ from Table A $= \underline{0.0968}$
The area between the limits $= 0.7802$

Answer:

78% of measurements meet the specification 12° ± 2°F.

This estimate of 78 percent meeting the specification 12°F ± 2°F compares favorably with the 81 percent we actually counted in Table 1.4.

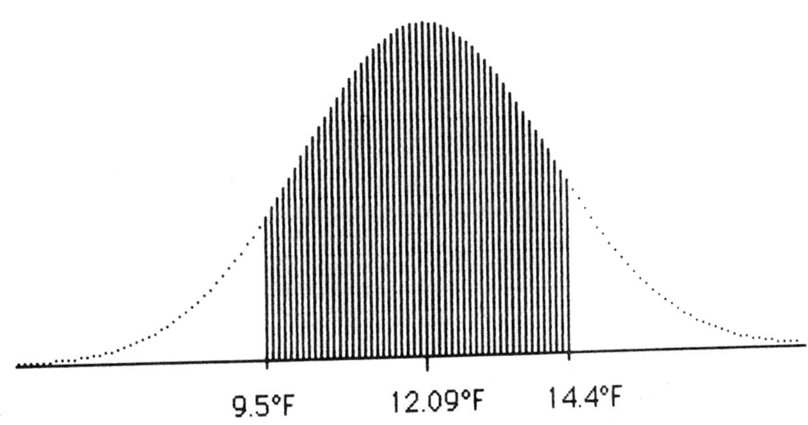

Figure 2.11 Area between 9.5°F and 14.4°F

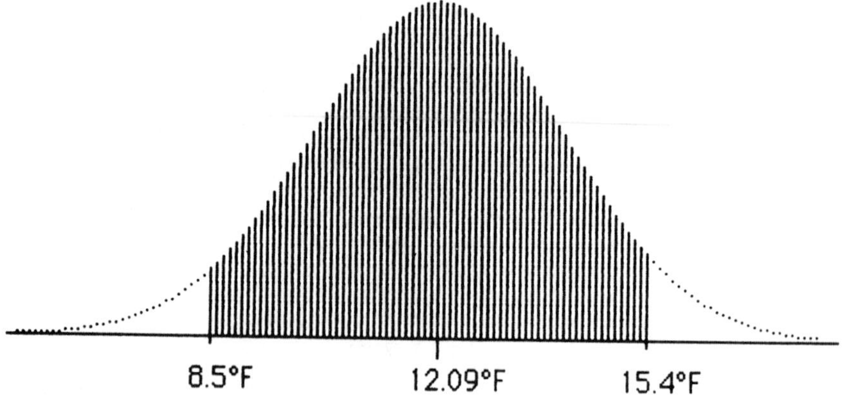

Figure 2.12 Area between 8.5°F and 15.4°F

Example 2-3

Given:

$\overline{X}$ = 12°F, σ = 1.985°F

Find:

Percentage of measurements meeting the specification 12°F ± 3°F

Solution:

Since the temperature values are rounded to the nearest integer, values between 8.5°F and 15.4°F meet these specifications (Figure 2.12).

$$Z = \frac{Xi - \overline{X}}{\sigma} = \frac{15.4 - 12.09}{1.985} = 1.67$$

$$Z = \frac{Xi - \overline{X}}{\sigma} = \frac{8.5 - 12.09}{1.985} = -1.81$$

The area from $-\infty$ to $+1.67\sigma$ from Table A = 0.9525
The area from $-\infty$ to -1.81σ from Table A = 0.0351
The area between the limits = 0.9174

Answer:

91.74% of measurements meet the specification 12°F ± 3°F.

Percentage of Production Out of Specification Limits

There are times when the available process will not have the capability to produce 100% of the product within the specification limits.

In these cases the process limits, $\overline{X} \pm 3\sigma$, are wider than the specification limits. In many cases itis desirable that all the reject parts be outside one of the limits rather than have a mix of too small and too big. Holes that are too small can be drilled or reamed a second time. Shafts that are too long can be machined again. By being able to calculate the percentage of production outside each specification limit, it is possible to determine the cost of scrap, the cost of reworking parts, and also to include the rework in production schedules. When the process limits, $\bar{X} \pm 3\sigma$, are greater than the specification limits, production can be held inside one of the specifications by setting the 3σ process limit equal to that specification limit and calculating the new process mean, $\bar{X}$.

The procedures for calculating the percentage production outside each specification limit and for calculating a mean, $\bar{X}$, which will make a process limit equal to a specification limit, are as follows:

- Calculate the process 3σ limits, $\bar{X} \pm 3\sigma$.
- Sketch process distribution and specification limits.
- Calculate the area out of the specification limits.

For areas to the left of the lower specification limit:

1. Calculate the number of standard deviations between the mean and the lower specification limit.
2. Get from Table A the area from $-\infty$ to the lower specification limit.

For areas to the right of the upper specification limit:

1. Calculate the number of standard deviations between the mean and the upper specification limit.
2. Get from Table A the area from $-\infty$ to the upper specification limit.
3. Subtract the area from $-\infty$ to the upper specification limit from 1 (the total area under the curve).

- To keep production above the lower specification, the new mean can be calculated from $\bar{X}$ = lower specification $+ 3\sigma$.
- To keep production below the upper specification, the new mean can be calculated from $\bar{X}$ = upper specification $- 3\sigma$.

Example 2-4

Problem:

A machining process produces shafts with an average diameter of 0.752 inch. The standard deviation has been calculated to be 0.008 inch. The specification calls for shaft measurements to be 0.750 ± 0.020 inch. Shafts that are too large can easily be reground. Shafts that are too small are scrap. What percentage of current production is scrap? What percentage can be reworked? If the average diameter can be shifted without changing the standard deviation, what new average diameter would eliminate scrap? What percentage of shafts would require rework with this new average diameter?

Given:

$\bar{X} = 0.752$ inch, $\sigma = 0.008$ inch

Specification = 0.750 ± 0.020 inch

Find:

a. Process limits.

b. % of production smaller than the lower specification (scrap).

c. % of production larger than the upper specification (rework).

d. New $\bar{X}$ at which scrap is minimized ($0.750 - 0.020 = \bar{X} - 3\sigma$).

e. % of production requiring rework with the new mean.

Solution:

a.

► Calculate the process 3σ limits, $\bar{X} \pm 3\sigma$.

Lower process limit $= \bar{X} - 3\sigma = 0.752 - 3 \times 0.008 = 0.728$ in.

Upper process limit $= \bar{X} + 3\sigma = 0.752 + 3 \times 0.008 = 0.776$ in.

► Sketch process distribution and specification limits (Figure 2.13)

► Calculate the area out of the specification limits.

b. For areas to the left of the lower specification limit:

1. Calculate the number of standard deviations between the mean and the lower specification limit.

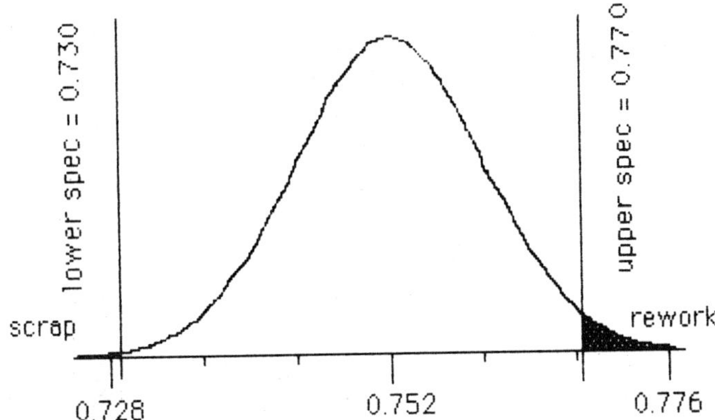

Figure 2.13 Process capability and specifications

2. Get from Table A the area from $-\infty$ to the lower specification limit.

$$Z = \frac{Xi - \bar{X}}{\sigma}$$

$$Z = \frac{0.730 - 0.752}{0.008} = -2.75$$

The area from $-\infty$ to -2.75σ from Table A = 0.0030, therefore 0.30% of the shaft diameters are below the lower specification.

c. For areas to the right of the upper specification limit:

1. Calculate the number of standard deviations between the mean and the upper specification limit.
2. Get from Table A the area from $-\infty$ to the upper specification limit.
3. Subtract the area from $-\infty$ to the upper specification limit from 1 (the total area under the curve).

$$Z = \frac{Xi - \bar{X}}{\sigma}$$

$$Z = \frac{0.770 - 0.752}{0.008} = 2.25$$

The area from $-\infty$ to 2.25σ from TAble A = 0.9878

Since 0.9878 is the area below the upper specification and the total area under the curve is 1.0000, then 1.0000 − 0.9878 is the area above the upper specification.

$$\begin{array}{r} 1.0000 \\ -\underline{0.9878} \\ 0.0122 \end{array}$$

1.22% of the shaft diameters are larger than the upper specification.

When the process limits, $\bar{X} \pm 3\sigma$, are greater than the specification limits, production can be held inside one of the specifications by setting the 3σ process limit equal to that specification and calculating the new process mean, $\bar{X}$.

► To keep production above the lower specification, the new mean can be calculated from $\bar{X}$ = lower specification + 3σ.

► To keep production below the upper specification, the new mean can be calculated from $\bar{X}$ = upper specification − 3σ.

d. To eliminate scrap, $\bar{X}$ = lower specification + 3σ.

$$\bar{X} = 730 + 3 \times 0.008 = .754 \text{ inch}$$

$$Z = \frac{Xi - \bar{X}}{\sigma}$$

$$Z = \frac{0.770 - 0.754}{0.008} = 2.0$$

e. The area from $-\infty$ to 2.0σ from Table A = 0.9773.

$$\begin{array}{r} 1.0000 \\ -\underline{0.9773} \\ 0.0227 \end{array}$$

2.27% of the shaft diameters would be larger than the upper specification and require rework.

Answer:

a. Process limits are 0.728 inch and 0.776 inch (Figure 2.14).

b. % < 0.750 − 0.020 inch (scrap) = 0.30%

c. % > 0.750 + 0.020 inch (rework) = 1.22%

d. New $\bar{X}$ at which scrap is eliminated = 0.754 inch

e. % rework at new $\bar{X}$ = 2.27%

Example 2-5

Problem:

The contained weight of a certain product is labeled as 500 grams. Measurement data shows the actual mean to be 505.2 grams and the standard deviation to be 1.2 grams. Assuming that the distribution is approximately normal, what are the process 3σ limits? To what value may the mean be lowered so that the lower process limit is 500.0 grams.

Given:

$\bar{X}$ = 505.2 grams, σ = 1.2 grams (Figure 2.15)

Find:

a. Process limits

b. New $\bar{X}$ so that $\bar{X} - 3\sigma = 500$ grams

Solution:

Process limits = $\bar{X} \pm 3\sigma$

Lower limit = $\bar{X} - 3\sigma$ = 505.2 grams − 3 × 1.2 grams = 501.6 grams

Upper limit = $\bar{X} + 3\sigma$ = 505.2 grams + 3 × 1.2 grams = 508.8 grams

New $\bar{X}$ = 500.0 grams + 3σ = 500 grams + 3 × 1.2 grams = 503.6 grams

New Upper limit = $\bar{X} + 3\sigma$ = 503.6 grams × 1.2 grams = 507.2 grams

Answer:

Process limits are 501.6 grams and 508.8 grams (Figure 2.16). New $\bar{X}$ = 503.6 grams

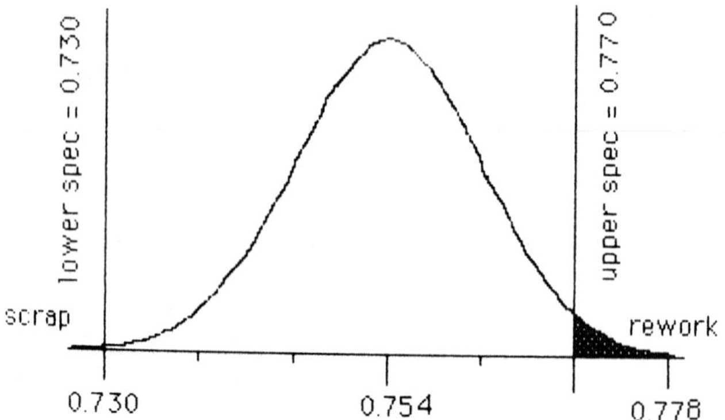

Figure 2.14 Process capability and specifications

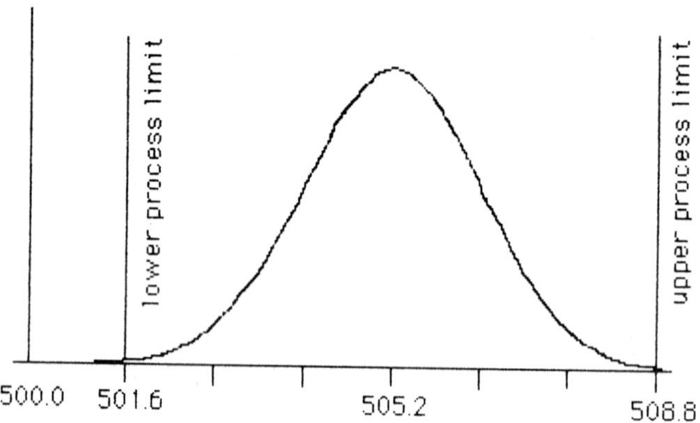

Figure 2.15 Process limits and labeled quantity

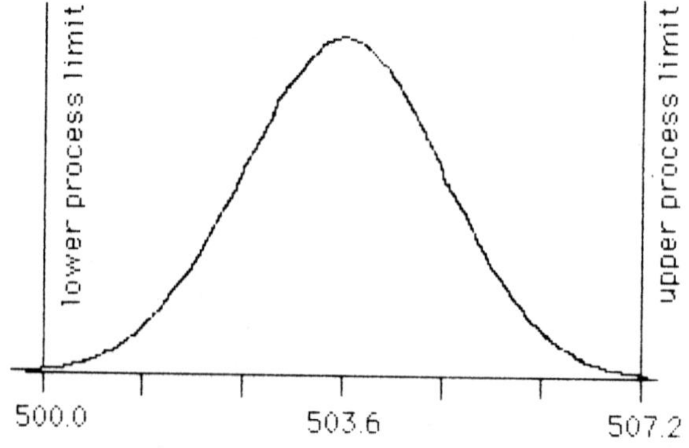

Figure 2.16 New process limits

Typical uses for a CMM include quick inspection of the first piece produced by a numerical control machine; sampling of production runs; maintenance inspection of dies, jigs, and fixtures; and acceptance inspection of machine parts. A CMM can check dimensional and geometric accuracy of everything from engine blocks and circuit boards to sheet metal parts and TV picture tubes. The speed and flexibility of CMM allows inspection to be part of the production process rather than reject-oriented.

Coordinate Measuring Machines have many advantages over the traditional measurement processes.

- One-time setup—Unlike surface plate techniques in which an inspector must make multiple setups to measure all axes of a part, a CMM setup consists of locating the part on the work table.
- No human error or variances—Although blatant measurement errors are few, an individual will get varying measurements for the same dimension and there will be a measurement variance between measurements taken by different people.
- Flexibility—The CMM can measure virtually any part within its size range without special gauging devices. Its memory can store measurement routines for many parts, making it possible to measure a variety of parts without changing setups or checking drawings between parts.
- Digital readouts, printed copy, computer records—CMMs are too expensive to have one at each production machine, but the ability to immediately print out measurement data that can be provided to the production worker is the next best thing. The CMM can record measurement data for management use.
- Geometric calculations—The CMM can make the geometric calculations to compare the relative location of all points on a part with the specifications.
- Statistical calculations and reports—The CMM can produce all the control charts necessary to monitor the production process without transcription of data.
- Computer aided design integration—The top of the line CMMs can receive measurement directions directly from the part design computer.

Chapter Review

Keywords

Normal curve The curve made by plotting the normal distribution. It has a mean of one and a standard deviation of one. It extends from $-\infty$ to $+\infty$. The total area under the curve is one.

Capability limits Upper and lower values of a process computed symmetrically around the mean. At least 99.73 percent of the process values fall between these limits.

Formulas

$$Z = \frac{Xi - X}{\sigma}$$

Z = Number of standard deviations, σ, between the mean, $\overline{X}$, and value x_i.

Process limits $= \overline{X} \pm 3\sigma$

Problems

2-1 A packaging operation is to provide 2 pounds, 32 ounces of a product in each container. After the process had stabilized, 100 packages were checked, and the results are shown in the frequency distribution in Table 2.3.

a. Compute $\overline{X}$ and σ

b. Compute the process limits, $\overline{X} \pm 3\sigma$.

c. What percentage of the containers would have less than 32.00 ounces?

d. What new process mean would place the lower process limit at 32.00 ounces?

Table 2.3 Packaging Operation Distribution

Weight	*Frequency*	*Weight*	*Frequency*
32.05	1	32.11	19
32.06	0	32.12	12
32.07	6	32.13	6
32.08	9	32.14	2
32.09	25	32.15	2
32.10	18		

2-2 The diameter of one hundred gear blanks was measured and the results are shown in the frequency distribution in Table 2.4.

a. Compute $\overline{X}$ and σ

b. Compute the process limits, $\overline{X} \pm 3\sigma$.

c. What percentage of production would meet the specification of 17.110 ± .030 inches?

d. If only those blanks larger than the upper specification can be reworked, what process mean would you recommend?

2-3 What percentage of a normal distribution would you expect to fall outside the limits $\overline{X} \pm 2.6\sigma$?

2-4 What measurements of the distribution shown in Problem 2-1 would be the $\overline{X} \pm 2.8\sigma$ limits.

2-5 What measurements of the distribution shown in Problem 2-2 would be the $\overline{X} \pm 2.9\sigma$ limits.

2-6 An approximately normal frequency distribution has a mean of 27.3 and a standard deviation of 1.2.

a. What are the process $\overline{X} \pm 3\sigma$ limits?

b. What percentage of the measurements would be expected to be above 25.1?

c. What percentage of the measurements would be expected to be below 27.2?

d. What percentage of the measurements would be expected to be between 26.9 and 28.0?

Table 2.4 Gear Blank Distribution

Diameter	*Frequency*	*Diameter*	*Frequency*
17.06	1	17.11	24
17.07	2	17.12	9
17.08	6	17.13	5
17.09	26	17.14	2
17.10	25		

2-7 The mean modulus of rupture for a large number of ceramic parts was found to be 7850 lb/in^2. The standard deviation is 880 lb/in^2 and the distribution is approximately normal.

a. What are the process $\bar{X} \pm 3\sigma$ limits?

b. What percentage of the measurements would be expected to be above 9000 lb/in^2?

c. What percentage of the measurements would be expected to be below 6000 lb/in^2?

d. What percentage of the measurements would be expected to be between 5000 lb/in^2 and 8000 lb/in^2?

2-8 A shipment of 5000 resistors has been received from a supplier. A random sample of 250 of these has been tested. The distribution is approximately normal with a mean of 76.7 ohms and a standard deviation of 3.9 ohms.

a. What are the process $\bar{X} \pm 3\sigma$ limits?

b. What percentage of the resistors would be expected to meet the specification of 75 ± 10 ohms?

c. What percentage of the resistors would be expected to be below 65 ohms?

d. What percentage of the resistors would be expected to be above 85 ohms?

2-9 A shipment of 2000 electric heaters has been received from a supplier. A random sample of 200 of these has been tested. The distribution is approximately normal with a mean output of 1100 watts and a standard deviation of 100 watts.

a. What are the process $\bar{X} \pm 3\sigma$ limits?

b. What percentage of the heaters would be expected to meet the specification of at least 1000 watts?

c. What percentage of the heaters would be expected to be below 950 watts?

d. What percentage of the heaters would be expected to be above 1200 watts?

3 Sample Data

This chapter covers the use of sample data to estimate the distribution of the process capability. It includes the characteristics of sample data.

Objectives

After completing this chapter you will be able to:

- calculate a sample mean, range, and standard deviation.
- from a group of samples, calculate the average sample mean, average sample range, average sample standard deviation, and the standard deviation of sample means.
- estimate the process standard deviation from the sample data.
- estimate the 3σ process limits from sample data.

Samples

A *sample* is some portion of a *population*. A *population* is the entire set. If we manufactured 1000 shafts and measured the diameter of each, those 1000 measurements would be the *population*. From them we could determine the process capability. If we measured the diameter of only 5 shafts, those 5 measurements would be a *sample*. In most manufacturing, it is neither economical nor necessary to measure every item produced. In statistical quality assurance, it is common to use sample measurements to estimate the distribution of the population and process capability.

Several samples are necessary to estimate the process capability. The measurement data of each sample (usually four of five measurements) will have a *sample mean*, $\bar{X}$, a *sample standard deviation*, σ, and a *range* R. The *range* is the difference between the largest measurement in the sample and the smallest measurement in the sample.

Groups of sample measurements also have distribution characteristics. Since each sample has a range, the group has an *average range*, $\bar{R}$. Since each sample has a standard deviation, the group has an *average sample standard deviation*, $\bar{\sigma}$. Since each sample has a mean, the group of samples has an *average sample mean*, $\bar{\bar{X}}$, and the distribution of the means has a *standard deviation of sample means*, $\sigma_{\bar{X}}$.

Quality Assurance Notation

We will be working with three groups of data: the process or population data, sample data, and data from groups of samples. All data groups have a mean, standard deviation, and a range. The symbols used in statistical quality assurance have been standardized. All symbols for characteristics of the population are followed by a prime, ′. The symbols we will use are shown in Table 3.1.

Table 3.1 Statistical Quality Assurance Symbols

POPULATION DATA		
Population mean	$\bar{X}'$	X bar prime
Population standard deviation	σ'	sigma prime
Population size	N	
SAMPLE DATA		
Sample mean	$\bar{X}$	X bar
Sample standard deviation	σ	sigma
Sample range	R	
Sample size	n	
GROUP OF SAMPLE DATA		
Average sample mean	$\bar{\bar{X}}$	X double bar
Average sample range	$\bar{R}$	R bar
Average sample standard deviation	$\bar{\sigma}$	sigma bar
Standard deviation of sample means	$\sigma_{\bar{X}}$	sigma X bar

Population Data

Table 3.2 contains the measurements of thermal switch trip temperatures discussed in Chapter 1. The 100 measurements are a population of size **N** = 100. In Chapter 1 we calculated $\bar{X}'$ to be 12.09°F and σ' to be 1.985°F.

Sample Data

The first 5 measurements in Table 3.2 are a sample of size $n = 5$.

The sample mean, $\bar{X}$, is the average value.

$$\bar{X} = \frac{\text{sample sum}}{n} = \frac{13 + 16 + 16 + 13 + 13}{5} = 14.2$$

The sample range, R, is the difference between the largest and smallest measurements in the sample.

$$R = \text{largest} - \text{smallest} = 16 - 13 = 3$$

Table 3.2 Trip Temperatures in °F.

13	***16***	***16***	***13***	***13***	9
12	15	8	12	13	14
11	6	15	11	15	13
12	***12***	***13***	***13***	***13***	11
17	14	11	14	13	15
11	11	13	15	9	10
11	12	11	16	11	10
13	***13***	***11***	***13***	***10***	13
11	10	8	11	11	17
10	12	13	10	10	15
14	***11***	***9***	***15***	***13***	12
11	14	12	10	12	14
14	12	11	11	12	13
11	***11***	***11***	***13***	***12***	11
12	10	9	12	10	13
12	***14***	***10***	***10***	***16***	12
11	11	14	14		

The sample standard deviation is the standard deviation of the five measurements.

$$\sigma = \sqrt{\frac{\Sigma(X - \bar{X})^2}{n}}$$

$$\sigma = \sqrt{\frac{(13-14.2)^2 + (16-14.2)^2 + (16-14.2)^2 + (13-14.2)^2 + (13-14.2)^2}{5}}$$

$$\sigma = 1.47$$

Data from Groups of Samples

Table 3.3 contains the data from the 6 samples shown in bold italics in Table 3.2.

The average sample mean of a group of samples is the mean of the means.

$$\bar{\bar{X}} = \frac{\Sigma \bar{X}}{\text{No. of samples}} = \frac{75.2}{6} = 12.53$$

The average sample range is the mean of the sample ranges.

$$\bar{R} = \frac{\Sigma R}{\text{No. of samples}} = \frac{21}{6} = 3.5$$

Table 3.3 Data from Six Samples

Sample Number	*Sample measurements (Sample size, n = 5)*					*Average* $\bar{X}$	*Range* R	*Sample Standard Deviation* σ
1	13	16	16	13	13	14.2	3	1.47
2	12	12	13	13	13	12.6	1	0.49
3	13	13	11	13	10	12.0	3	1.26
4	14	11	9	15	13	12.4	6	2.15
5	11	11	11	13	12	11.6	2	0.80
6	12	14	10	10	16	12.4	6	2.33
						$\Sigma\bar{X} = 75.2$	$\Sigma R = 21$	$\Sigma\sigma = 8.50$

The average sample standard deviation is the mean of the sample standard deviations.

$$\bar{\sigma} = \frac{\Sigma\sigma}{\text{No. of samples}} = \frac{8.50}{6} = 1.42$$

The standard deviation of the sample means is the standard deviation of the $\bar{X}$ values.

$$\sigma_{\bar{X}} = \sqrt{\frac{(14.2-12.53)^2+(12.6-12.53)^2+(12-12.53)^2+(12.4-12.53)^2+(11.6-12.53)^2+(12.4-12.53)^2}{6}} = 0.81$$

Example 3-1

Given:

The following data from six samples of five measurements each.

Sample	*Measurements*				
1	22.48	23.85	21.86	22.55	22.6
2	20.48	22.01	23.64	20.25	22.0
3	22.67	22.87	21.67	19.21	23.69
4	21.33	23.26	22.52	21.94	22.14
5	22.38	22.61	22.28	21.4	24.4
6	23.08	21.57	23.24	22.67	23.55

Find:

The average sample mean, the average sample range, the average sample standard deviation, and the standard deviation of the sample means.

Solution:

Table 3.4.

$\bar{R} = 2.795$

$\bar{\bar{X}} = 22.344$

$\bar{\sigma} = 0.956$

$\sigma_{\bar{X}} = 0.3988$

Table 3.4 Data from Six Samples

Sample Number	*Sample measurements (Sample size, n = 5)*					*Sample Range* R	*Sample Average* $\bar{X}$	*Sample Standard Deviation* σ
1	22.48	23.85	21.86	22.55	22.64	1.99	22.676	0.648
2	20.48	22.01	23.64	20.25	22.07	3.39	21.69	1.232
3	22.67	22.87	21.67	19.21	23.69	4.48	22.022	1.546
4	21.33	23.26	22.52	21.94	22.14	1.93	22.238	0.640
5	22.38	22.61	22.28	21.40	24.40	3	22.614	0.983
6	23.08	21.57	23.24	22.67	23.55	1.98	22.822	0.687
					TOTAL	16.77	134.062	5.736

Estimation of Population σ' from Sample Data

Statistical theory provides several relationships between sample data and the population from which the samples were taken where the population is approximately normal. It tells us that in the long run, the average of sample means, $\bar{\bar{X}}$, will equal the population mean, $\bar{X}'$. The standard deviation of the sample means will be the standard deviation of the population divided by the square root of the sample size. That is $\sigma_{\bar{X}} = \sigma'/\sqrt{n}$. If the sample size were four, the $\sigma_{\bar{X}}$ would be half the population σ'. The ratio of the population standard deviation and the average standard deviation of a group of samples is designated as c_2. $c_2 = \bar{\sigma}/\sigma'$. The ratio of the of the population standard deviation and the average of the sample ranges is designated as d_2. $d_2 = \bar{R}/\sigma'$ The values of c_2 and d_2 depend on the number of measurements in each sample. They are listed in Appendix A, Table B. We will use these relationships to estimate the population characteristics and the process limits where the population can be assumed to be approximately normal. The accuracy of the estimate will depend on the number of samples in the sample group.

Estimating Population Characteristics from Sample Data

In Table 3.2 we selected six samples from the thermal switch data given in Chapter 1. In that chapter we calculated the population mean, $\bar{X}'$, to be 12.09°F and the σ' to be 1.985°F. From Table 3.3 we calculated from the sample data that $\bar{\sigma} = 1.42$, $\bar{\bar{X}} = 12.53$, $\bar{R} = 3.5$, and $\sigma_{\bar{X}} = 0.81$.

Estimated Population Characteristics Using Sample Information

The population mean from $\bar{\bar{X}}$
$\bar{X}' = \bar{\bar{X}} = 12.53$

The population standard deviation from $\sigma_{\bar{X}}$
$\sigma' = {}_{\bar{X}}\sqrt{n} = 0.81\sqrt{5} = 1.81$

The population standard deviation from $\bar{\sigma}$
$\sigma' = \bar{\sigma}/c_2$

From Table B, c_2 for a sample of size 5 is 0.8407
$\sigma' = 1.42 / 0.8407 = 1.69$

The population standard deviation from $\bar{R}$
$\sigma' = \bar{R} / d_2$

From Table B, d_2 for a sample of size 5 is 2.326
$\sigma' = 3.5 / 2.326 = 1.5$

Now use Table 3.5 to compare the estimates with the actual values calculated in Chapter 1.

The 24.4% error in the estimation of σ' from the $\bar{R}$ is excessive but a group of six samples is not very large. A more common example would be to select approximately twenty samples from a production run of 500. Table 3.6 shows in italics the samples taken. The actual mean of the 500 numbers in Table 3.6 is 12.07. The actual standard deviation is 2.056.

Table 3.5 Estimates and Actual Values

Characteristic	*Actual*	*Estimate*	*From*	*% Error*
$\bar{X}'$	12.09	12.53	$\bar{\bar{X}}$	3.6
σ'	1.985	1.81	$\sigma_{\bar{X}}$	8.8
σ'	1.985	1.69	$\bar{\sigma}$	14.9
σ'	1.985	1.5	$\bar{R}$	24.4

Table 3.6 Production Run of 500

Measurements									
13	16	12	13	13	9	12	15	8	12
13	14	11	6	15	11	15	13	12	12
13	*13*	*13*	*11*	*17*	14	11	14	13	15
11	11	13	15	9	10	11	12	11	16
11	10	13	13	11	13	10	13	11	10
8	*11*	*11*	*17*	*10*	12	13	10	10	15
14	11	9	15	13	12	11	14	12	10
12	14	14	12	11	11	12	13	11	11
11	*13*	*12*	*11*	*12*	10	9	12	10	13
12	14	10	10	16	12	11	11	14	14
10	12	11	14	12	15	12	11	10	15
14	*11*	*9*	*12*	*15*	11	10	11	9	15
9	10	13	15	9	14	14	15		
10	12	11	14	12	11	13	11	13	11
12	*13*	*14*	*15*	*9*	9	10	12	15	11
14	14	14	14	11	11	10	10	12	11
18	*9*	*13*	*11*	*13*	11	11	12	12	11
10	*13*	*13*	*13*	*9*	14	11	11	11	15
14	11	12	13	12	14	14	14	10	12
14	14	12	16	14	11	9	13	12	14
15	*11*	*15*	*12*	*11*	12	9	11	12	10
15	12	10	14	11	13	13	13	14	17
11	*11*	*12*	*12*	*14*	13	13	13	12	15
13	*15*	*12*	*12*	*10*	11	11	12	11	11
14	12	14	13	11	13	11	12	16	10

Table 3.6 Continued

Measurements									
11	10	14	14	14	12	17	13	13	10
10	*9*	*12*	*12*	*12*	12	11	10	14	11
11	16	10	10	10	13	12	13	10	11
13	11	11	16	12	14	11	11	9	11
11	*11*	*14*	*14*	*15*	15	11	13	13	9
8	10	8	11	11	12	14	10	13	15
8	11	12	11	10	15	13	12	13	12
16	*12*	*13*	*11*	*10*	11	10	9	12	14
9	9	8	7	13	11	11	13	12	13
13	14	10	9	15	12	14	10	10	11
14	*12*	*14*	*11*	*10*	11	11	17	12	9
13	16	15	11	12	13	12	13	14	10
17	*10*	*16*	*13*	*14*	14	16	7	11	16
11	*12*	*14*	*11*	*8*	11	9	12	11	12
12	18	10	8	16	10	15	10	13	11
12	14	11	14	11	12	11	11	17	16
14	*12*	*12*	*8*	*13*	13	13	13	12	13
13	10	16	12	11	13	13	14	11	16
12	11	11	12	8	13	16	10	11	9
14	*14*	*15*	*13*	*9*	10	13	11	9	11
9	11	13	12	13	14	10	13	12	16
9	12	13	12	15	7	12	12	15	11
14	*13*	*11*	*13*	*13*	12	13	14	12	9
16	8	14	8	13	9	10	9	9	8
14	*12*	*9*	*12*	*12*	13	14	12	9	10

Table 3.7 Data from 20 Samples

Sample Number	*Sample measurement (Sample size, n = 5)*					*Sample Range* *R*	*Sample Average* $\bar{X}$	*Sample Standard Deviation* σ
sample1	13	13	13	11	17	6	13.4	1.960
sample2	8	11	11	17	10	9	11.4	3.007
sample3	11	13	12	11	12	2	11.8	0.748
sample4	14	11	9	15	13	6	12.4	2.154
sample5	12	13	14	15	9	6	12.6	2.059
sample6	18	9	13	11	13	9	12.8	2.993
sample7	10	13	13	13	9	4	11.6	1.744
sample8	15	11	15	12	11	4	12.8	1.833
sample9	11	11	12	12	14	3	12	1.095
sample10	13	15	12	12	10	5	12.4	1.625
sample11	10	9	12	12	12	3	11	1.265
sample12	11	11	14	14	15	4	13	1.673
sample13	16	12	13	11	10	6	12.4	2.059
sample14	14	12	14	11	10	4	12.2	1.600
sample15	17	10	16	13	14	7	14	2.449
sample16	11	12	14	11	8	6	11.2	1.939
sample17	14	14	15	13	9	6	13	2.098
sample18	10	13	13	13	9	4	11.6	1.744
sample19	14	13	11	13	13	3	12.8	0.980
sample20	14	12	9	12	12	5	11.8	1.600
				TOTAL		102	246.2	36.625

Now to calculate $\bar{R}$, $\bar{\sigma}$, $\bar{\bar{X}}$, and $\sigma_{\bar{X}}$ using the sample information from Table 3.7.

$\bar{R} = 5.1$

$\bar{\bar{X}} = 12.31$

$\bar{\sigma} = 1.83$

$\sigma_{\bar{X}} = 0.75$

We can now estimate $\bar{X}'$ and σ'.

The population mean from $\bar{\bar{X}}$
$\bar{X}' = \bar{\bar{X}} = 12.31$

The population standard deviation from $\sigma_{\bar{X}}$
$\sigma' = \sigma_{\bar{X}} \sqrt{n} = 0.75 \sqrt{5} = 1.68$

The population standard deviation from $\bar{\sigma}$
$\sigma' = \bar{\sigma} / c_2$

From Table B, c_2 for a sample of size 5 is 0.8407
$\sigma' = 1.83 / 0.8407 = 2.18$

The population standard deviation from $\bar{R}$
$\sigma' = \bar{R} / d_2$

From Table B, d_2 for a sample of size 5 is 2.326
$\sigma' = 5.1 / 2.326 = 2.19$

Now use Table 3.8 to compare the estimates with the actual values.

Estimates of σ' are usually made from $\bar{R}$ or $\bar{\sigma}$ because they are reasonably accurate and easy to calculate. Unless statistical calculators are available, the range is the easier to calculate than the standard deviation.

Table 3.8 Estimates and Actual Values

Characteristic	*Actual*	*Estimate*	*From*	*% Error*
$\bar{X}'$	12.07	12.31	$\bar{\bar{X}}$	2
σ'	2.056	1.68	$\sigma_{\bar{X}}$	18
σ'	2.056	2.18	$\bar{\sigma}$	6.5
σ'	2.056	2.19	$\bar{R}$	6.5

Table 3.9 Data from 20 Samples

sample1	32.39	33.48	31.89	32.44	32.51
sample2	30.79	32.01	33.31	30.6	32.05
sample3	32.53	32.69	31.73	29.77	33.35
sample4	31.46	33.01	32.41	31.96	32.11
sample5	32.31	32.48	32.23	31.52	33.92
sample6	32.86	31.65	32.99	32.53	33.24
sample7	31.72	31.8	32.43	33.3	30.72
sample8	31.23	31.51	31.93	31.75	33.79
sample9	31.65	31.23	32.27	32.34	31.54
sample10	32.47	31.39	32.59	31.46	31.34
sample11	30.6	31.41	31.71	33.94	31.37
sample12	32.1	32.58	31.14	31.2	33.14
sample13	32.74	31.44	30.77	33.03	32.49
sample14	32.13	31.57	32.64	32.12	31.29
sample15	31.9	32.87	32.71	31.81	31.62
sample16	31.4	32.15	32.58	31.45	31.52
sample17	31.53	32.48	32.05	31.72	31.93
sample18	31.2	30.92	32.09	31.25	32.43
sample19	31.95	32.71	31.36	31.06	33.43
sample20	32.07	31.76	31.58	32.83	32.66

Estimation of Process Limits from Sample Data

The estimation of process limits from sample data is almost identical to the calculation of process limits from population data. The difference is that the popuation mean and standard deviation must first be estimated from the sample data. The procedure is as follows:

- Calculate the mean and range or standard deviation of each sample.
- Calculate the average sample mean and average sample range or average standard deviation.
- Estimate the population mean and standard deviation.
- Calculate the process limits as $\bar{X}' \pm 3\,\sigma'$

Example 3-2

Given:

The measurement data from twenty samples of five measurements each, shown in Table 3.9.

Find:

The process $\bar{X}' \pm 3\,\sigma'$ limits.

Solution:

► Calculate the mean and range or standard deviation of each sample. (For the purpose of the example, both the range and standard deviation are shown in Table 3.10.)

► Calculate the average sample mean and average sample range or average standard deviation.

$$\bar{R} = \frac{38.38}{20} = 1.919$$

$$\bar{\bar{X}} = \frac{641.022}{20} = 32.0511$$

$$\bar{\sigma} = \frac{13.798}{20} = 0.6899$$

► Estimate the population mean and standard deviation.

The population standard deviation from $\bar{\sigma}$

$$\sigma' = \bar{\sigma} / c_2$$

Table 3.10 Data from 20 Samples

Sample Number	*Sample measurements (Sample size, n = 5)*					*Sample Range* R	*Sample Average* $\bar{X}$	*Sample Standard Deviation* σ
sample1	32.39	33.48	31.89	32.44	32.51	1.59	32.542	0.518
sample2	30.79	32.01	33.31	30.6	32.05	2.71	31.752	0.983
sample3	32.53	32.69	31.73	29.77	33.35	3.58	32.014	1.235
sample4	31.46	33.01	32.41	31.96	32.11	1.55	32.19	0.512
sample5	32.31	32.48	32.23	31.52	33.92	2.4	32.492	0.786
sample6	32.86	31.65	32.99	32.53	33.24	1.59	32.654	0.552
sample7	31.72	31.8	32.43	33.3	30.72	2.58	31.994	0.852
sample8	31.23	31.51	31.93	31.75	33.79	2.56	32.042	0.905
sample9	31.65	31.23	32.27	32.34	31.54	1.11	31.806	0.431
sample10	32.47	31.39	32.59	31.46	31.34	1.25	31.85	0.558
sample11	30.6	31.41	31.71	33.94	31.37	3.34	31.806	1.128
sample12	32.1	32.58	31.14	31.2	33.14	2	32.032	0.777
sample13	32.74	31.44	30.77	33.03	32.49	2.26	32.094	0.852
sample14	32.13	31.57	32.64	32.12	31.29	1.35	31.95	0.473
sample15	31.9	32.87	32.71	31.81	31.62	1.25	32.182	0.507
sample16	31.4	32.15	32.58	31.45	31.52	1.18	31.82	0.467
sample17	31.53	32.48	32.05	31.72	31.93	0.95	31.942	0.323
sample18	31.2	30.92	32.09	31.25	32.43	1.51	31.578	0.578
sample19	31.95	32.71	31.36	31.06	33.43	2.37	32.102	0.871
sample20	32.07	31.76	31.58	32.83	32.66	1.25	32.18	0.490
					TOTAL	38.38	641.022	13.798

From Table B, c_2 for a sample of size 5 is 0.8407

$\sigma' = 0.6899 / 0.8407 = 0.821$

The population standard deviation from $\bar{R}$

$\sigma' = \bar{R} / d_2$

From Table B, d_2 for a sample of size 5 is 2.326

$\sigma' = 1.919 / 2.326 = 0.825$

► Calculate the process $\bar{X}' \pm 3\sigma'$ limits.

Limits calculated from σ' are:

Upper limit $= \bar{X}' + 3\sigma' = 32.0511 + 3 \times 0.821$
Upper limit $= 34.514$

Lower limit $= \bar{X}' - 3\sigma' = 32.0511 - 3 \times 0.821$
Lower limit $= 29.588$

Limits calculated from $\bar{R}$ are:

Upper limit $= \bar{X}' + 3\sigma' = 32.0511 + 3 \times 0.825$
Upper limit $= 34.526$

Lower limit $= \bar{X}' - 3\sigma' = 32.0511 - 3 \times 0.825$
Lower limit $= 29.576$

Chapter Review

Key Words

Population (universe) The entire set of data of interest in a statistical analysis.

Sample A portion of a population.

Population Data

Population mean	$\bar{X}'$	X bar prime
Population standard deviation	σ'	sigma prime
Population size	N	

Sample Data		
Sample mean	$\bar{X}$	X bar
Sample standard deviation	σ	sigma
Sample range	R	
Sample size	n	

Group of Sample Data		
Average sample mean	$\bar{\bar{X}}$	X double bar
Average sample range	$\bar{R}$	R bar
Average sample standard deviation	$\bar{\sigma}$	sigma bar
Standard deviation of sample means	$\sigma_{\bar{X}}$	sigma X bar

Formulas

$$\bar{X} = \frac{\text{sample sum}}{n}$$

$$R = \text{largest} - \text{smallest}$$

$$\bar{\bar{X}} = \frac{\Sigma \bar{X}}{\text{No. of samples}}$$

$$\bar{R} = \frac{\Sigma R}{\text{No. of samples}}$$

$$\bar{\sigma} = \frac{\Sigma \sigma}{\text{No. of samples}}$$

$$\sigma' = \sigma_{\bar{X}} \sqrt{n}$$

$$\sigma' = \bar{\sigma} / c_2$$

$$\sigma' = \bar{R} / d_2$$

Problems

3-1 Table 3.11 shows the measurements taken in twenty samples of five measurements each.

Find:

a. The process mean.
b. The standard deviation of sample 5.
c. The average sample range.
d. The estimated population standard deviation.
e. The process three sigma limits, assuming the process to be normally distributed.

Table 3.11 Measurements from 20 Samples (Problem 3-1)

1	48.53	50.04	47.85	48.60	48.70
2	46.33	48.01	49.81	46.07	48.07
3	48.74	48.95	47.63	44.93	49.86
4	47.26	49.39	48.57	47.94	48.15
5	48.42	48.67	48.31	47.34	50.64
6	49.19	47.52	49.36	48.73	49.70
7	47.62	47.72	48.60	49.79	46.25
8	46.95	47.32	47.91	47.65	50.46
9	47.51	46.94	48.38	48.47	47.36
10	48.64	47.17	48.81	47.26	47.09
11	46.07	47.19	47.60	50.66	47.13
12	48.14	48.79	46.82	46.90	49.57
13	49.02	47.24	46.31	49.41	48.68
14	48.18	47.41	48.88	48.16	47.03
15	47.86	49.20	48.98	47.74	47.48
16	47.18	48.21	48.79	47.24	47.35
17	47.35	48.66	48.07	47.62	47.90
18	46.90	46.52	48.13	46.96	48.60
19	47.94	48.98	47.12	46.71	49.96
20	48.10	47.68	47.42	49.15	48.91

3-2 Table 3.12 lists the measurements taken in twenty samples of five measurements each.

Find:

a. The process mean.
b. The standard deviation of sample 5.
c. The average sample range.
d. The estimated population standard deviation.
e. The process three sigma limits, assuming the process to be normally distributed.

Table 3.12 Measurements from 20 Samples (Problem 3-2)

1	3.05	3.19	2.99	3.05	3.06
2	2.85	3.00	3.16	2.82	3.01
3	3.07	3.09	2.97	2.72	3.17
4	2.93	3.13	3.05	2.99	3.01
5	3.04	3.06	3.03	2.94	3.24
6	3.11	2.96	3.12	3.07	3.15
7	2.97	2.97	3.05	3.16	2.84
8	2.90	2.94	2.99	2.97	3.22
9	2.96	2.90	3.03	3.04	2.94
10	3.06	2.92	3.07	2.93	2.92
11	2.82	2.93	2.96	3.24	2.92
12	3.01	3.07	2.89	2.90	3.14
13	3.09	2.93	2.85	3.13	3.06
14	3.02	2.95	3.08	3.01	2.91
15	2.99	3.11	3.09	2.98	2.95
16	2.93	3.02	3.07	2.93	2.94
17	2.94	3.06	3.01	2.97	2.99
18	2.90	2.87	3.01	2.91	3.05
19	2.99	3.09	2.92	2.88	3.18
20	3.01	2.97	2.95	3.10	3.08

3-3 Table 3.13 lists the measurements taken in twenty samples of five measurements each.

Find:

a. The process mean.
b. The standard deviation of sample 1.
c. The average sample range.
d. The estimated population standard deviation.
e. The process three sigma limits, assuming the process to be normally distributed.

Table 3.13 Measurements from 20 Samples (Problem 3-3)

1	145.45	149.55	143.59	145.65	145.91
2	139.45	144.03	148.93	138.74	144.20
3	14601	146.60	143.00	135.63	149.08
4	141.98	147.78	145.55	143.83	144.41
5	145.15	145.82	144.85	142.21	151.20
6	147.24	142.70	147.71	146.00	148.65
7	142.96	143.24	145.62	148.88	139.22
8	141.13	142.16	143.75	143.05	150.70
9	142.68	141.10	145.02	145.28	142.26
10	145.75	141.73	146.20	141.98	141.52
11	138.74	141.79	142.92	151.26	141.63
12	144.38	146.16	140.78	140.99	148.29
13	146.78	141.92	139.38	147.86	145.85
14	144.50	142.40	146.41	144.45	141.34
15	143.61	147.26	146.66	143.29	142.59
16	141.76	144.56	146.16	141.93	142.22
17	142.23	145.79	144.19	142.97	143.74
18	141.00	139.97	144.34	141.17	145.62
19	143.83	146.67	141.61	140.49	149.35
20	144.27	143.12	142.41	147.12	146.48

3-4 Table 3.14 lists the measurements taken in twenty samples of five measurements each.

Find:

a. The process mean.
b. The standard deviation of sample 6.
c. The average sample range.
d. The estimated population standard deviation.
e. The process three sigma limits, assuming the process to be normally distributed.

Table 3.14 Measurements from 20 Samples (Problem 3-4)

1	1.02	1.09	0.99	1.03	1.03
2	0.92	1.00	1.08	0.91	1.00
3	1.03	1.04	0.98	0.86	1.08
4	0.97	1.06	1.03	1.00	1.01
5	1.02	1.03	1.01	0.97	1.12
6	1.05	0.98	1.06	1.03	1.08
7	0.98	0.99	1.03	1.08	0.92
8	0.95	0.97	1.00	0.98	1.11
9	0.98	0.95	1.02	1.02	0.97
10	1.03	0.96	1.04	0.97	0.96
11	0.91	0.96	0.98	1.12	0.96
12	1.01	1.04	0.95	0.95	1.07
13	1.05	0.97	0.92	1.06	1.03
14	1.01	0.97	1.04	1.01	0.96
15	0.99	1.05	1.04	0.99	0.98
16	0.96	1.01	1.04	0.97	0.97
17	0.97	1.03	1.00	0.98	1.00
18	0.95	0.93	1.01	0.95	1.03
19	1.00	1.04	0.96	0.94	1.09
20	1.00	0.99	0.97	1.05	1.04

3-5 In a laboratory test of cement blocks, 20 samples of four blocks were subjected to compression tests. The mean breaking stress was 3210 pounds per square inch. The average range for the 20 samples was 210 pounds per square inch. Estimate sigma prime and calculate the process limits. Assume the population to be normally distributed.

3-6 Twenty-five samples of five measurements each were taken of the diameter of a casting. The mean diameter was 21.6 inches. The average sample standard deviation was 0.31 inches. Assume the population to be normally distributed. Can this process be used to meet a specification of 20 in. ± 1.5 in.?

3-7 A container-filling process must meet a specification of a minimum of 26 ounces in each container. Thirty samples of six measurements produced $\bar{\bar{X}} = 26.21$ and $\bar{R} = 0.18$. Assume the population to be normally distributed. Can this process be used?

3-8 A resistor is manufactured to meet the specification 5 Ω ± 0.01Ω. Twenty-five samples of four measurements showed $\Sigma\bar{X} = 125.05\Omega$ and $\Sigma R = 0.171\Omega$. Assume the population to be normally distributed. Can this process be used ?

3-9 Thirty samples of five measurements were taken of the diameter of a shaft. The sample data was calculated to be $\Sigma\bar{X} = 33.05$ inches and $\Sigma R = 0.35$ inch. Assume the population to be normally distributed and calculate the process limits.

3-10 The length of a shaft must meet the specification 10.25 ± 0.05 inch. The data from twenty-five samples of 5 measurements was calculated to be $\Sigma\bar{X} = 255.225$ inches and $\Sigma R = 0.275$ inch. Assume the population to be normally distributed. What are the process limits?

4 Control Charts

In the previous chapters we learned to use population and sample data to determine the process capability and to compare the process capability with specifications. In this chapter we will learn to use sample data collected during production to assure that the process does not vary without detection.

Objectives

After completing this chapter you will be able to:

- use control charts.
- develop control charts for sample averages.
- develop control charts for sample ranges.
- develop control charts for sample standard deviations.
- interpret control charts.
- use control charts in decision making.
- use control charts in complex processes.

Sample Mean Distribution

Thus far in this text we have worked with populations that are normally distributed. It is a statistical fact that even though a population is not normal, the distribution of sample means will be very close to normal for sample sizes of four or more. This fact will allow us to detect a shift in the population by observing a shift in the distribution of sample means. Given that the distribution of sample means is normal, 99.73% of the sample means should be between $\bar{\bar{X}} \pm 3\sigma_{\bar{X}}$.

Control Charts

A control chart is a graph of sample data plotted over time (Figure 4.1). It includes a centerline, an upper control limit (UCL) and a lower control limit (LCL). The centerline is the mean value of the sample data plotted such as $\bar{\bar{X}}$. The upper and lower control limits are the 3 sigma limits of the sample data such as $\bar{\bar{X}} + 3\,\sigma_{\bar{X}}$ and $\bar{\bar{X}} - 3\,\sigma_{\bar{X}}$. Figure 4.1 shows the format of a typical control chart. When the sample data vary normally within the UCL and LCL, the process is said to be *in control*. That is, the variability of measurements is due to chance causes, not a change in the process. When sample data is out of a control limit, the process is said to be *out of control*. That is, something probably has caused the process to change. The purpose of the control chart is to detect this change during production.

Once the centerline and limits are plotted, samples are collected periodically during production. On an $\bar{X}$ chart each sample mean is plotted on the chart in the order in which the samples were taken.

Figure 4.2 shows an $\bar{X}$ chart of the sample averages calculated in Example 3-2. Note that the sample averages are plotted from left to right in the order in which they were taken.

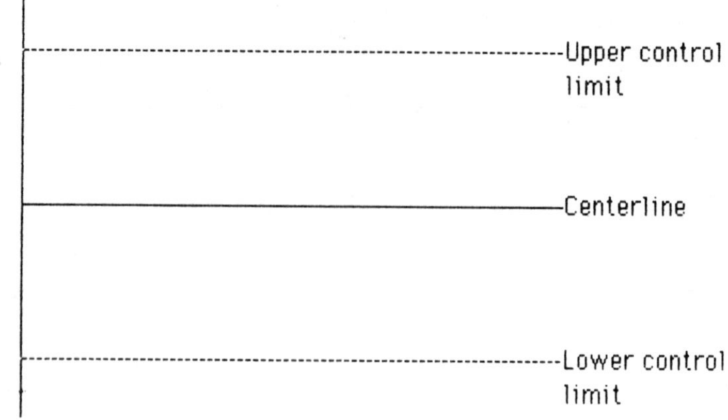

Figure 4.1 Control Chart Format

The $\bar{X}$ chart will show the shift of the population $\bar{X}'$, but the mean is not the only measure of a distribution. The mean could remain unchanged while the standard deviation changes. To assure control, a second chart is necessary to detect shifts in the standard deviation. This second chart is either a chart of sample ranges, R chart, or a chart of sample standard deviations, σ chart. The $\bar{X}$ chart is plotted above the second chart as shown in Figure 4.3, which displays the sample means and ranges from Example 3-2.

Use of Control Charts

Control charts serve two major purposes: they show changes in the process distribution; they show when changes in sample values are probably caused by chance rather than a process change. By early detection of process changes, it is usually possible to make corrections before any of the items produced exceed the specifications. This avoids scrap or rework. Knowing when high or low sample values are within the control limits prevents unnecessary production delays caused by looking for nonexistent problems.

Changes in production can be detected from the charts in two ways. Of the plotted measurements 99.7% should be within the control limits. Sample values outside the control limits are reason for concern. Since the centerline is the mean of the normally distributed sample values, about an equal number of the measurements should be on each side of the centerline. Getting seven successive measurements on one side has the same probability of chance causes as getting seven successive heads or tails when flipping a coin.

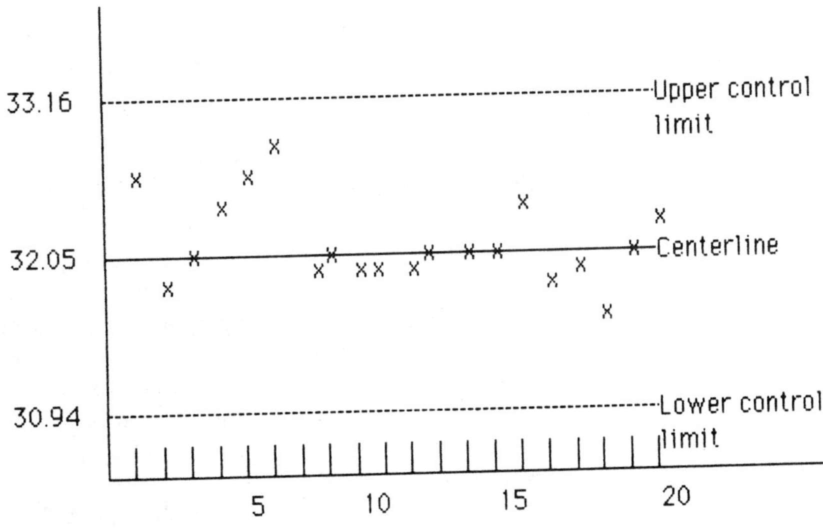

Figure 4.2 $\bar{X}$ CHART

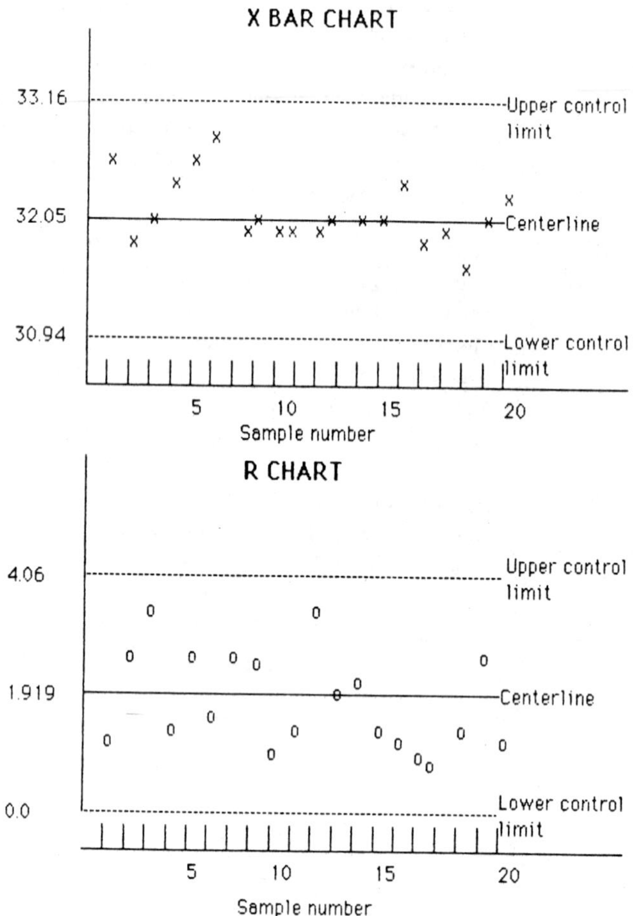

Figure 4.3 Process Control Charts

How much concern the sample results generate is an economic decision. It depends on the relationship between the process limits and the specifications, and the cost of reworking or scrapping the item produced. If the process limits are well within the specification, a clear indication of process change (more than 2% of sample values out of the control limits) would occur before the specifications are violated. If the process limits are close to the specifications and scrap or rework is undesirable, a single sample value out of the control limits or seven successive values on the same side of the centerline justifies increased sampling. In Chapter Six we will discuss the determination of sample size and sample frequency.

Developing Control Charts

In order to be statistically reliable, at least 100 measurements are required to calculate process limits and control chart limits. If earlier estimates are desired, they may be calculated with fewer measurements but should be updated when more data is available.

X CHART

The $\bar{X}$ CHART is a chronological graph of sample means. The centerline is the average of the sample means, $\bar{\bar{X}}$. The upper control limit is $\bar{\bar{X}} + 3\sigma_{\bar{X}}$. The lower control limit is $\bar{\bar{X}} - 3\sigma_{\bar{X}}$. The standard deviation of sample means can be calculated by the equation $\sigma_{\bar{X}} = \sigma' / \sqrt{n}$. If σ' is not known, it can be estimated from either the average sample range, $\bar{R}$, or the average sample standard deviation, $\bar{\sigma}$.

Depending on the information available, the $\bar{X}$ CHART limits can be calculated as:

$$\bar{\bar{X}} \pm 3\sigma_{\bar{X}} = \bar{\bar{X}} \pm \frac{3\,\sigma'}{\sqrt{n}}$$

$$\bar{\bar{X}} \pm 3\sigma_{\bar{X}} = \bar{\bar{X}} \pm \frac{3\bar{R}}{d_2\sqrt{n}}$$

$$\bar{\bar{X}} \pm 3\sigma_{\bar{X}} = \bar{\bar{X}} \pm \frac{3\bar{\sigma}}{c_2\sqrt{n}}$$

To shorten the calculation of control chart limits, the multipliers preceding σ', $\bar{R}$, and $\bar{\sigma}$ have been tabulated in Appendix A, Table B, for sample sizes of 2 to 10 measurements. Using the multipliers, the three $\bar{X}$ CHART limits can be calculated as:

$$\bar{\bar{X}} \pm 3\sigma_{\bar{X}} = \bar{X}' \pm A\,\sigma'$$

$$\bar{\bar{X}} \pm 3\sigma_{\bar{X}} = \bar{\bar{X}} \pm A_2\bar{R}$$

$$\bar{\bar{X}} \pm 3\sigma_{\bar{X}} = \bar{\bar{X}} \pm A_1\bar{\sigma}$$

In Example 3-2 we calculated $\bar{\bar{X}} = 32.05$, $\bar{R} = 1.919$. The sample size is five. Table B shows the value of A_2 for a sample of five to be 0.577.

The upper and lower control limits for the $\bar{X}$ chart in Figure 4.2 are calculated as follows:

$$UCL_{\bar{X}} = \bar{\bar{X}} + A_2\bar{R} = 32.05 + .577 \times 1.919 = 33.16$$

$$LCL_{\bar{X}} = \bar{\bar{X}} - A_2\bar{R} = 32.05 - .577 \times 1.919 = 30.94$$

R CHART

The centerline of the R CHART is $\bar{R}$. If the chart is developed from population data, the centerline is $d_2\sigma'$.

Using sample range data, the upper control limit is $UCL_R = D_4 \bar{R}$. The lower control limit is $LCL_R = D_3 \bar{R}$.

Using population data, the upper control limit is $UCL_R = D_2\sigma'$. The lower control limit is $LCL_R = D_1 \sigma'$.

The values of D_1, D_2, D_3, and D_4 are listed in Table B for sample sizes from two to ten. For samples of six or fewer measurements, the lower limit of the R chart is zero.

The upper and lower control limits for the R chart in Figure 4.3 were calculated as follows:

$$UCL_R = D_4 \bar{R} = 2.115 \times 1.919 = 4.06$$

$$LCL_R = D_3 \bar{R} = 0 \times 1.919 = 0$$

σ CHART

The centerline of the σ CHART is $\bar{\sigma}$. If the chart is developed from population data, the centerline is $c_2\sigma'$.

Using sample data, the upper control limit is $UCL_\sigma = B_4 \bar{\sigma}$. The lower control limit is $UCL_\sigma = B_3 \bar{\sigma}$.

Using population data, the upper control limit is $UCL_\sigma = B_2 \sigma'$. The lower control limit is $UCL_\sigma = B_1 \sigma'$.

The values of B_1, B_2, B_3, and B_4 are listed in Table B for sample sizes from two to ten. For samples of six or fewer measurements, the lower limit of the σ chart is zero.

A sumary of the equations for developing control charts for variables is in Table 4.1.

The procedure for the development of control charts is as follows:

- Tabulate sample data, preferably a minimum of 100 measurements.
- Calculate each sample mean and range or standard deviation.
- Calculate the sample group $\bar{\bar{X}}$ and $\bar{R}$, or $\bar{\bar{X}}$ and $\bar{\sigma}$.
- Select the proper equations from Table 4.1 and calculate the chart centerline, UCL, and LCL.
- Graph the control charts as a pair with the $\bar{X}$ chart on top.
- Plot the sample data to assure that the process was in control when the data were collected.

Table 4.1 Equations for Developing Control Charts

Data Available	$\bar{X}$ Chart	*R Chart*	σ *Chart*
Population	Centerline $= \bar{X}'$	Centerline $= d_2\sigma$	Centerline $= c_2\sigma'$
$\bar{X}'$ and σ'	$UCL_{\bar{X}} = \bar{X}' + A\sigma'$	$UCL_R = D_2\sigma'$	$UCL_\sigma = B_2\sigma'$
	$LCL_{\bar{X}} = \bar{X} - A\sigma'$	$LCL_R = D_1\sigma'$	$LCL_\sigma = B_1\sigma$
Sample	Centerline $= \bar{\bar{X}}$	Centerline $= \bar{R}$	
$\bar{\bar{X}}$ and $\bar{R}$	$ULC_{\bar{X}} = \bar{\bar{X}} + A_2\bar{R}$	$UCL_R = D_4\bar{R}$	
	$LCL_{\bar{X}} = \bar{\bar{X}} - A_2\bar{R}$	$LCL_R = D_3\bar{R}$	
Sample	Centerline $= \bar{\bar{X}}$		Centerline $= \bar{\sigma}$
$\bar{\bar{X}}$ and $\bar{\sigma}$	$UCL_{\bar{X}} = \bar{\bar{X}} + A_1\bar{\sigma}$		$UCL_\sigma = B_4\bar{\sigma}$
	$LCL_{\bar{X}} = \bar{\bar{X}} - A_1\bar{\sigma}$		$LCL_\sigma = B_3\bar{\sigma}$

Example 4-1

Given:

The sample measurements shown in Table 4.2.

Find:

Develop control charts for $\bar{X}$ and σ.

Solution:

► Tabulate sample data.

► Calculate each sample mean and range or standard deviation, (Table 4.2).

► Calculate the sample group $\bar{\bar{X}}$ and $\bar{R}$, or $\bar{\bar{X}}$ and $\bar{\sigma}$.

$$\bar{\bar{X}} = \frac{641.022}{20} = 32.0511$$

$$\bar{\sigma} = \frac{13.798}{10} = 0.6899$$

► Select the proper equations from Table 4.1 and calculate the chart centerline, UCL, and LCL.

From Table 4.1:

Data Available	**$\bar{X}$ Chart**	**σ Chart**
Sample $\bar{\bar{X}}$ *and* $\bar{\sigma}$	Centerline $= \bar{\bar{X}}$	Centerline $= \bar{\sigma}$
	$UCL_{\bar{X}} = \bar{\bar{X}} + A_1\bar{\sigma}$	$UCL_{\sigma} = B_4\bar{\sigma}$
	$LCL_{\bar{X}} = \bar{\bar{X}} - A_1\bar{\sigma}$	$LCL_{\sigma} = B_3\bar{\sigma}$

From Table B, for sample size n = 5:

$A_1 = 1.596$, $B_4 = 2.089$, $B_3 = 0$

$\bar{X}$ Chart

Centerline $= \bar{\bar{X}} = 32.05$

$UCL_{\bar{X}} = \bar{\bar{X}} + A_1\bar{\sigma} = 32.05 + 1.596 \times 0.6899$ $UCL_{\bar{X}} = 33.15$

$LCL_{\bar{X}} = \bar{\bar{X}} - A_1\bar{\sigma} = 32.05 - 1.596 \times 0.6899$ $LCL_{\bar{X}} = 30.95$

Table 4.2 Sample Measurements (Example 4-1)

Sample Number	*Sample measurements (Sample size, n = 5)*					*Sample Average* $\bar{X}$	*Sample Standard Deviation* σ
sample 1	32.39	33.48	31.89	32.44	32.51	32.542	0.518
sample 2	30.79	32.01	33.31	30.60	32.05	31.752	0.983
sample 3	32.53	32.69	31.73	29.77	33.35	32.014	1.235
sample 4	31.46	33.01	32.41	31.96	32.11	32.190	0.512
sample 5	32.31	32.48	32.23	31.52	33.92	32.492	0.786
sample 6	32.86	31.65	32.99	32.53	33.24	32.654	0.552
sample 7	31.72	31.80	32.43	33.30	30.72	31.994	0.852
sample 8	31.23	31.51	31.93	31.75	33.79	32.042	0.905
sample 9	31.65	31.23	32.27	32.34	31.54	31.806	0.431
sample 10	32.47	31.39	32.59	31.46	31.34	31.850	0.558
sample 11	30.60	31.41	31.71	33.94	31.37	31.806	1.128
sample 12	32.10	32.58	31.14	31.20	33.14	32.032	0.777
sample 13	32.74	31.44	30.77	33.03	32.49	32.094	0.852
sample 14	32.13	31.57	32.64	32.12	31.29	31.950	0.473
sample 15	31.90	32.87	32.71	31.81	31.62	32.182	0.507
sample 16	31.40	32.15	32.58	31.45	31.52	31.820	0.467
sample 17	31.53	32.48	32.05	31.72	31.93	31.942	0.323
sample 18	31.20	30.92	32.09	31.25	32.43	31.578	0.578
sample 19	31.95	32.71	31.36	31.06	33.43	32.102	0.871
sample 20	32.07	31.76	31.58	32.83	32.66	32.180	0.490
					TOTAL	641.022	13.798

σ Chart

Centerline $= \bar{\sigma} = 0.6899$

$UCL_\sigma = B_4\,\bar{\sigma} = 2.089 \times 0.6899$ $\qquad UCL_\sigma = 1.44$

$LCL_\sigma = B_3\,\bar{\sigma} = 0 \times 0.6899$ $\qquad LCL_\sigma = 0$

► Graph the control charts as a pair with the $\overline{X}$ chart on top.

► Plot the sample data to assure that the process was in control when the data were collected. Control charts cannot be developed from data showing the process is out of control.

Figure 4.4 shows the developed charts with the sample data plotted.

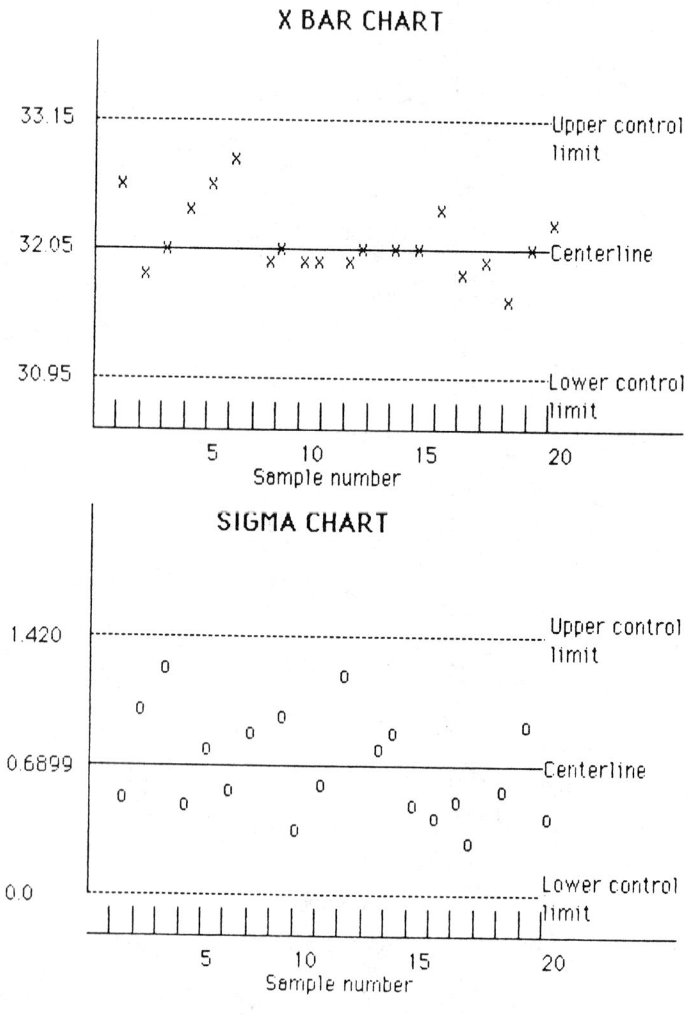

Figure 4.4 Solution to Example 4-1.

Interpreting of Control Charts

Control charts are a graphic display of the output of a process over time. While the process is in control (only chance variations) the points plotted should be normally distributed. If they fall out of the limits, the process has changed. Even though the plotted points lie within the limits, a change has probably occurred when seven consecutive points fall on the same side of the centerline. If a process is sensitive to cyclical changes such as temperature or pressure, this can be reflected in control charts by the movement of sample means.

Control charts show what a process is doing. A thorough understanding of the process is necessary to find the cause of a process going out of control.

Control Charts in Decision Making

In many production organizations, similar work is done many times. Lathes may turn out many different parts in the course of a year, but the turning process has limited variations. The historical record contained in the control chart files shows the previous records of many man, material, machine combinations. This is useful scheduling information for future work. In the current environment of rapidly changing tool materials and treatments, it is not as easy as it once was for the manager to remember the capability of each available combination. As new tools and processes are used, the control chart provides an analytical basis for evaluating successes and failures.

Control Charts in Complex Process

In many processes the output is determined by several inputs of varying importance. In the manufacture of a thin chemical film on an acetate sheet, for example, the thickness of the film was determined by the temperature of the chemical, the pressure at which the chemical was fed, and the velocity of the sheet. The customer specifications had only a minimum film thickness. Using control charts only on the film thickness would insure a salable product but perhaps not the most profitable one. Film material may be wasted by producing an excess thickness.

In setting up such a process for the first time, control charts should initially be established on all the inputs as well as the product. These charts show the natural limits on each input. With an examination of the result that input changes have on the product, it will be possible to determine the mean values for temperature, pressure, and velocity which produce a useful product with the minimun expense. Control charts on any input which is found to have little effect or low probability of changing could be discontinued.

In evaluating complex processes, control charts provide easy statistical analysis of measurement data.

Chapter Review

Keywords

In Control Process A process in which the variability is due to chance causes.

Out of Control Process A process in which variability is due to a process change.

Control Chart Limits, UCL, LCL The three sigma distance from the mean value of a sample. UCL_R is the upper control limit of an R chart and is three standard deviations above $\bar{R}$. LCL_R is the lower control limit of an R chart and is three standard deviation below $\bar{R}$.

Formulas

Control Chart Limits

Data Available	$\bar{X}$ Chart	*R Chart*	*σ Chart*
Population	Centerline $= \bar{X}'$	Centerline $= d_2\sigma$	Centerline $= c_2\sigma'$
$\bar{X}'$ and σ'	$UCL_{\bar{X}} = \bar{X}' + A\sigma'$	$UCL_R = D_2\sigma'$	$UCL_\sigma = B_2\sigma'$
	$LCL_{\bar{X}} = \bar{X} - A\sigma'$	$LCL_R = D_1\sigma'$	$LCL_\sigma = B_1\sigma$
Sample	Centerline $= \bar{\bar{X}}$	Centerline $= \bar{R}$	
$\bar{\bar{X}}$ and $\bar{R}$	$ULC_{\bar{X}} = \bar{\bar{X}} + A_2\bar{R}$	$UCL_R = D_4\bar{R}$	
	$LCL_{\bar{X}} = \bar{\bar{X}} - A_2\bar{R}$	$LCL_R = D_3\bar{R}$	
Sample	Centerline $= \bar{\bar{X}}$		Centerline $= \bar{\sigma}$
$\bar{\bar{X}}$ and $\bar{\sigma}$	$UCL_{\bar{X}} = \bar{\bar{X}} + A_1\bar{\sigma}$		$UCL_\sigma = B_4\bar{\sigma}$
	$LCL_{\bar{X}} = \bar{\bar{X}} - A_1\bar{\sigma}$		$LCL_\sigma = B_3\bar{\sigma}$

Problems

4-1 Using the data in Table 3.5, prepare $\bar{X}$ and R charts. Plot the sample data.

4-2 Using the data in Problem 3-1, prepare $\bar{X}$ and R charts. Plot the sample data.

4-3 Using the data in Table 3.5, prepare $\bar{X}$ and σ charts. Plot the sample data.

4-4 Using the data in Problem 3-2, prepare $\bar{X}$ and σ charts. Plot the sample data.

4-5 Using the data in Problem 3-3, prepare $\bar{X}$ and R charts. Plot the sample data.

4-6 Using the data in Problem 3-4, prepare $\bar{X}$ and R charts. Plot the sample data.

4-7 The first 25 samples of a shaft length gave a mean of 11.25 inches and an average range of 0.10 inch. Calculate the centerline, UCL, and LCL for $\bar{X}$ and R charts to be used with a sample size of six. What percentage of production will meet a specification of 11.25 ± .05 inch.

4-8 A filling operation must put at least 1.50 ounces of a product into a container. Current control chart data show a mean of 1.58 ounces with an average range of .020 ounce for samples of five measurements. What are the current process limits? Assume the process distribution is normal. The mean can be shifted without changing the range. What new mean can be used to meet the specification and minimize cost?

4-9 The first 150 items of a production run have been measured. The mean of the measurements is 0.671 inch. The standard deviation of the 150 measurements is 0.003 inch. It is desired to use samples of five measurements to control the production. Calculate the centerline, UCL, and LCL for the $\bar{X}$ and R charts to be used.

4-10 The first 25 samples of four measurements each show the samples to have a mean of 4.41 inches and an average sample standard deviation of 0.020 inch. Calculate the centerline, UCL, and LCL for the $\bar{X}$ and O charts to be used.

4-11 The first 20 samples of five measurements each show the samples to have a mean of 2.70 inches and an average sample standard deviation of 0.010 inch. Calculate the centerline, UCL, and LCL for the $\bar{X}$ and σ charts to be used. Calculate the process limits and the percentage of production that would meet a specification of 2.70 ± 0.03 inch.

5 Probability

In the previous chapters we discussed the distribution of measurement data. In the following chapters we will discuss the distribution of attribute data, things measured or counted in integers. In order to understand the distribution of attribute data, it is necessary to understand the laws of probability.

Objectives

After completing this chapter you will be able to:

- calculate the probability of simple events.
- calculate the probability of compound events.
- calculate the number of combinations of possible events.
- calculate the probability that a sample will contain a certain number of bad items.

Probability Theory

The *American Heritage Dictionary* defines probability as "a number expressing the likelihood of occurrence of an event, such as the ratio of the number of experimental results that would produce the event to the total number of events considered possible." In the quality assurance area it is thought of as the relative frequency of a certain result in the long run.

The normal distribution is a probability distribution. When we calculated that 95% of production would be between certain limits, we also calculated that an individual piece selected from that production would have a 95% probability of being between those limits.

The area under the normal curve is 1 (one). Probability theory states that the total probability of all events considered possible is 1 (one). This total of all possible events is called the *probability space*, S. In tossing a coin and not letting it remain on end, there are only two possible events, Head and Tail. This probability space is:

$$S = \{H,T\}$$

The probability space for rolling a single die is:

$$S = \{1,2,3,4,5,6\}$$

It is impossible for an event that is not in the probability space to occur. There is always some probability that an event in the probability space will occur.

Simple Probability

An event, which will be designated by a capital letter, must be one of the events in S. In tossing a coin, we could designate event A as getting a tail. Then:

$$A = \{T\}$$

which is one of the events in S = {H,T}.

The probability of a single event is equal to the number of ways that event could occur divided by the total number of events in the probability space. The probability of event A is designated as P(A).

The probability of getting a tail in the single flip of a coin is:

$$P(A) = \frac{1}{2}$$

There is only one tail in the probability space {H,T}, so the numerator is one. There are two events in the probability space {H,T}, so the denominator is two.

If we designate event A as getting a 2 in a roll of a single die, the probability is:

$$P(A) = \frac{1}{6}$$

There is only one 2 in the probability space {1,2,3,4,5,6}, so the numerator is one. There are six events in the probability space {1,2,3,4,5,6} so the denominator is six.

Example 5-1

Given:

A box contains three red balls and seven blue balls.

Find:

a. The probability of getting a red ball on a single draw.
b. The probability of getting a blue ball on a single draw.

Solution:

The probability space contains ten balls,

$$S = \{RRRBBBBBBB\}$$

a. Event A is getting one of the three red balls.

$$A = \{R\}$$

Answer:

$$\text{P(A)} = \frac{3}{10}$$

b. Event B is getting one of the seven blue balls.

$$B = \{B\}$$

$$P(B) = \frac{7}{10}$$

In this example, the probability of getting a blue ball could have been calculated in another manner. There are only two types of balls in the probability space. The probability of getting a blue ball is the same as the probability of *not* getting a red ball.

Since only events in the probability space, S, can occur, P(S) = 1. The total probability of all the simple events in the probability space, S, must be one. Therefore:

$$P(A) + P(B) = 1$$

and

$$P(B) = 1 - P(A)$$

$$P(B) = 1 - \frac{3}{10} = \frac{7}{10}$$

Example 5-2

Given:

A normal deck of 52 playing cards.

Find:

a. The probability of picking any ace in a single draw.
b. The probability of picking the ace of clubs in a single draw.
c. The probability of picking any club in a single draw.
d. The probability of not getting a diamond in a single draw.

Solution:

Since the deck has 52 cards, S = {52 cards}. Four of the cards are aces, one is the ace of clubs; thirteen of the cards are clubs and thirteen are diamonds. If 13 cards are diamonds, then 39 are not diamonds.

Event A = {any ace}
Event B = {the ace of clubs}
Event C = {any club}
Event D = {not a diamond}
S = {52 cards}

Answer:

a. $P(A) = \frac{4}{52}$

b. $P(B) = \frac{1}{52}$

c. $P(C) = \frac{13}{52}$

d. $P(D) = \frac{39}{52}$ *or*

$$P(D) = 1 - \frac{13}{52} = \frac{39}{52}$$

Compound Probability

We will define compound probability as probability concerning two or more events. The probability of drawing an ace or a club, or the probability of successive events, such as getting two heads in two coin flips, are examples. There are two major types of compound probability.

- The probability of two or more events, such as the probability of getting a head on the first toss *and* getting a second head on the second toss.
- The probability of getting one of several events, such as drawing an ace *or* a king from a card deck.

The Probability of Two or More Events

The probability of two events, A and B, is equal to the probability of event A times the probability of event B given that event A has occurred.

$$P(A \text{ and } B) = P(A) \times P(B|A)$$

The vertical line in the expression P(B|A) stands for the word *given*. The multiplication of the probabilities is logical. Probabilities are less than one or equal to one. Multiplication of fractions produces a value smaller than either multiplier. The probability of two selected events should be lower than the probability of either. If logic fails, remember when you see *and*, multiply.

The reason for the probability of B given A, P(B|A), is also logical. In some cases, getting A changes the probability of B.

Example 5-3

Given:

A box contains three red balls and seven blue balls. Selected balls are not returned to the container.

Find:

a. The probability of getting two red balls on successive draws.
b. The probability of getting three blue balls on successive draws.

Solution:

Let events be:

A = {first red ball}

B = {second red ball}

C = {first blue ball}

D = {second blue ball}

E = {third blue ball}

S = {10 balls; 3 R, 7B}

The probability of the first red ball will be the same as in Example 5-2:

$$P(A) = \frac{3}{10}$$

Since the first ball is not returned to the container, the probability of the second red ball will be different. Given that the first ball picked was red, the container now contains only nine balls, two of which are red. Therefore,

$$P(B|A) = \frac{2}{9}$$

$$P(A \text{ and } B) = P(A) \times P(B|A)$$

$$P(A \text{ and } B) = \frac{3}{10} \times \frac{2}{9}$$

Answer a:

$$P(A \text{ and } B) = 0.067$$

$$P(C \text{ and } D \text{ and } E) = P(C) \times P(D|C) \times P(E|C \,\&\, D)$$

$$P(C \text{ and } D \text{ and } E) = \frac{7}{10} \times \frac{6}{9} \times \frac{5}{8}$$

Answer b:

$$P(C \text{ and } D \text{ and } E) = 0.29$$

By not replacing each ball before selecting the next, we made the second and third events dependent on the preceding events. If we replace each ball after observing its color, the second and third selections are independent of preceding events. The probability of picking a blue ball remains unchanged no matter how many blue balls have already been picked.

Example 5-4

Given:

A box contains three red balls and seven blue balls. Selected balls are returned to the container before the next selection.

Find:

a. The probability of getting two red balls on successivew draws.
b. The probability of getting three blue balls on successive draws.

Solution:

Let events be:

A = {first red ball}

B = {second red ball}

C = {first blue ball}

D = {second blue ball}

E = {third blue ball}

S = {10 balls; 3R, 7B}

$$P(A \text{ and } B) = P(A) \times P(B|A)$$

$$P(A \text{ and } B) = \frac{3}{10} \times \frac{3}{10}$$

Answer a:

$$P(A \text{ and } B) = 0.09$$

$$P(C \text{ and } D \text{ and } E) = P(C) \times P(D|C) \times P(E|C \,\&\, D)$$

$$P(C \text{ and } D \text{ and } E) = \frac{7}{10} \times \frac{7}{10} \times \frac{7}{10}$$

Answer b:

$$P(C \text{ and } D \text{ and } E) = 0.34$$

Some events are independent by their nature. In flipping a coin, each flip is independent. The probability of a head on the first flip is 0.5. The coin has no sense of history so the probability of getting a head on the seventh flip is 0.5 whether or not the preceding six flips were heads. The probability of seven successive heads, however, would be:

P(seven successive heads) $= 0.5 \times 0.5 \times 0.5 \times 0.5 \times 0.5 \times 0.5 \times 0.5$

P(seven successive heads) $\times\ 0.0078$

The Probability of One of Several Events

The probability of getting one of several events, A or B, is equal to the probability of even A plus the probability of event B minus the probability of getting both A and B, event AB, if they are not *mutually exclusive*.

$$P(A \text{ or } B) = P(A) + P(B) - P(AB)$$

The addition of the probabilities is logical. There should be a higher probability of getting either of two events than getting a specific event. If logic fails, when you see *or*, add. Two events are mutually exclusive if getting one of them precludes getting the other. In the coin flip, heads and tails are mutually exclusive because you cannot get both on a single flip. In a card deck, getting an ace or a club are not mutually exclusive events because there is an ace of clubs.

Example 5-5

Given:

A normal deck of 52 playing cards.

Find:

a. The probability of an ace or diamond in a single draw.
b. The probability of an ace or a king in a single draw.

Solution:

Let A = {ace}
D = {diamond}
K = {king}
S = {52 cards}

a. This part can be solved in two ways. We can count all the events that satisfy the requirement and treat it as a simple probability problem. There are thirteen diamonds and three aces that are not diamonds. Sixteen of the fifty-two cards are A or D.

$$P(A \text{ or } D) = \frac{16}{52}$$

Using this formula, we get:

$$P(A \text{ or } D) = P(A) + P(D) - P(AD)$$

$$P(A \text{ or } D) = \frac{4}{52} + \frac{13}{52} - \frac{1}{52}$$

Answer a:

$$\text{P(A or } D) = \frac{16}{52}$$

You can see that the first part of the formula counted the ace of diamonds twice: first as an ace and then as a diamond. The expression − P(AD) corrected the double counting.

b. $P(A \text{ or } K) = P(A) + P(K) - P(AK)$

$$P(A \text{ or } K) = \frac{4}{52} + \frac{4}{52} - 0$$

Answer b:

$$P(A \text{ or } K) = \frac{8}{52}$$

In part b, the expression $-P(AK)$ is zero because a single card cannot be both an ace and a king.

Example 5-6

Given:

A container has three red balls, four blue balls, and five green balls.

Find:

a. The probability of getting a red ball and then a green ball in two draws without replacing the drawn balls.
b. The probability of getting either a green or blue ball on the first draw.
c. The probability of getting a red ball and then a green ball in two draws if the first ball is returned to the container before the second draw.

Solution:

Let R = {red ball}
G = {green ball}
B = {blue ball}
S = {12 balls, 3R, 4B, 5G}

a. $P(R \text{ and } G) = P(R) \times P(G|R)$

$$P(R \text{ and } G) = \frac{3}{12} \times \frac{5}{11}$$

Answer a:

$$P(R \text{ and } G) = 0.114$$

$$P(G \text{ or } B) = P(G) + P(B) - P(GB)$$

$$P(G \text{ or } B) = \frac{5}{12} + \frac{4}{12} - 0$$

Answer b:

$$P(G \text{ or } B) = \frac{9}{12}$$

$$P(R \text{ and } G) = P(R) \times P(G|R)$$

$$P(R \text{ and } G) = \frac{3}{12} \times \frac{5}{12}$$

Answer c:

$$P(R \text{ and } G) = 0.104$$

The Probability of Combined Events

In many cases the events to be examined are a combination of *and* and *or* situations.

Example 5-7

Given:

A coin to be flipped.

Find:

The probability of seven successive heads or tails.

Solution:

Let H = {head}
T = {tail}
S = {H, T}

Since the two events are independent and mutually exclusive, we can leave out the "given" and the − P(HT) and get:

$$P(7H \text{ or } 7T) = P(7H) + P(7T)$$

$$P(7H) = P(H) \times P(H) \times P(H) \times P(H) \times P(H) \times P(H) \times P(H)$$

$$P(7H) = 0.5^7 = 0.0078$$

$$P(7T) = P(T) \times P(T) \times P(T) \times P(T) \times P(T) \times P(T) \times P(T)$$

$$P(7T) = 0.5^7 = 0.0078$$

$$P(7H \text{ or } 7T) = P(7H) + P(7T)$$

$$P(7H \text{ or } 7T) = 0.0078 + 0.0078$$

Answer:

$$P(7H \text{ or } 7T) = 0.0156$$

This is also the probability that seven successive sample data points will be on the same side of a control chart centerline if there has been no change in the process.

The procedure for the solution of probability problems is as follows:

- Define the events in the probability space.
- Define the events for which the probability is being calculated.
- Determine which events are combined as *and* events and *or* events.
- Determine if the events are independent and mutually exclusive.
- If both *and* events and *or* events are involved, calculate the probability of the *and* events first.

Example 5-8

Given:

One lot of 100 capacitors is to be sampled. The lot has 10 bad capacitors.

Find:

The probability that a sample of five will have exactly one bad capacitor.

Solution:

► Define the events in the probability space.

S = {90 good cap. + 10 bad cap.}

► Define the events for which the probability is being calculated.

Let G = {good cap.}
B = {bad cap.}
A = {The probability that a sample of five will have exactly one bad capacitor.}

► Determine which events are combined as *and* events and *or* events.

There are five possible samples that have exactly one bad capacitor. The bad capacitor could be selected first, second, third, fourth, *or* fifth. In each sample, one must be bad *and* the remaining four good. The five possible samples with exactly one bad capacitor are:

B *and* G *and* G *and* G *and* G

or

G *and* B *and* G *and* G *and* G

or

G *and* G *and* B *and* G *and* G

or

G *and* G *and* G *and* B *and* G

or

G *and* G *and* G *and* G *and* B

► Determine if the events are independent and mutually exclusive.

A capacitor cannot be both good and bad. There is no event BG. Assuming that the first inspected capacitor is not returned to the lot before next one is selected, the probabilities are dependent. The probability on each selection changes as the population of 100 is decreased.

► If both *and* events and *or* events are involved, calculate the probability of the *and* events first.

$$P(B \text{ and } G \text{ and } G \text{ and } G \text{ and } G) = \frac{10}{100} \times \frac{90}{99} \times \frac{89}{98} \times \frac{88}{97} \times \frac{87}{96} = 0.0679$$

$$P(G \text{ and } B \text{ and } G \text{ and } G \text{ and } G) = \frac{90}{100} \times \frac{10}{99} \times \frac{89}{98} \times \frac{88}{97} \times \frac{87}{96} = 0.0679$$

$$P(G \text{ and } G \text{ and } B \text{ and } G \text{ and } G) = \frac{90}{100} \times \frac{89}{99} \times \frac{10}{98} \times \frac{88}{97} \times \frac{87}{96} = 0.0679$$

$$P(G \text{ and } G \text{ and } G \text{ and } B \text{ and } G) = \frac{90}{100} \times \frac{89}{99} \times \frac{88}{98} \times \frac{10}{97} \times \frac{87}{96} = 0.0679$$

$$P(G\ and\ G\ and\ G\ and\ G\ and\ B) = \frac{90}{100} \times \frac{89}{99} \times \frac{88}{98} \times \frac{87}{97} \times \frac{10}{96} = 0.0679$$

$$P(A) = P(B\ and\ G\ and\ G\ and\ G\ and\ G) + P(G\ and\ B\ and\ G\ and\ G\ and\ G) + P(G\ and\ G\ and\ B\ and\ G\ and\ G) + P(G\ and\ G\ and\ G\ and\ B\ and\ G) + P(G\ and\ G\ and\ G\ and\ G\ and\ B)$$

$$P(A) = 0.0679 + 0.0679 + 0.0679 + 0.0679 + 0.0679$$

Answer:

$$P(A) = 0.3395$$

Notice that the probability of each of the five combinations that contain exactly one bad capacitor is the same. In each of the five calculations, the denominators are identical. The numerators are not identical but have the same five multipliers. Only the order is different and that does not change the value of the multiplication. When the event is an exact number of bad items in a given sample size, we could calculate the probability of that event by calculating the probability of one combination and multiplying that value by the number of possible combinations.

In Example 5-8 we could have calculated:

$$P(A) = 5 \times P(B\ and\ G\ and\ G\ and\ G\ and\ G)$$

$$P(A) = 5 \times 0.0679$$

$$P(A) = 0.3395$$

Combinations

In Example 5-8 it is easy to visualize the number of sample groups that contain one bad item. To visualize the number of samples of ten items that contain three bad parts is more difficult and time consuming. There is a formula that will calculate the number of different sets of r objects that can be in a group of n objects.

$$C_r^n = \frac{n!}{r!(n-r)!}$$

The term $n!$ is read an n factorial and is the product of the first n integers. $4! = 4 \times 3 \times 2 \times 1 = 1 = 24$. By definition, $0! = 1$.

To calculate the possible number of samples of five that can have exactly one bad capacitor, we would have:

$r = 1$

$n = 5$

$$C^n_r = \frac{n!}{r!(n-r)!}$$

$$C^5_1 = \frac{5!}{1!(5-1)!} = \frac{5 \times 4 \times 3 \times 2 \times 1}{1(4 \times 3 \times 2 \times 1)}$$

$$C^5_1 = 5$$

To calculate the number of samples of ten items that contain three bad parts, we would have:

$$C^n_r = \frac{n!}{r!(n-r)!}$$

$$C^{10}_3 = \frac{10!}{3!(10-3)!} = \frac{10 \times 9 \times 8 \times 7 \times \ldots \times 1}{3 \times 2 \times 1(7 \times \ldots \times 1)} = \frac{10 \times 9 \times 8}{3 \times 2 \times 1}$$

$$C^{10}_3 = 120$$

In quality assurance we are more concerned with a range of bad items in a sample than with an exact number. We would be concerned that a sample taken from a shipment of a certain expected quality would have as *many as five* bad items rather than the probability that it would have *exactly five* bad items.

Example 5-9

Given:

One lot of 100 capacitors is to be sampled. The lot has 10 bad capacitors.

Find:

The probability that a sample of five will have two or less bad capacitors.

Solution:

A sample of five can only have:

exactly 5 bad capacitors

exactly 4 bad capacitors

exactly 3 bad capacitors

exactly 2 bad capacitors

exactly 1 bad capacitor

exactly 0 bad capacitors

Two or less is 2, 1, or 0
P(Two or less) = P(2) + P(1) + P(0)

One of the ways a sample can have exactly two bad capacitors is B B G G G

There are C^5_2 possible combinations of two bad and three good capacitors. The probability of exactly two bad capacitors in a sample of five is:

$$P(2) = C^5_2\, P(B \text{ and } B \text{ and } G \text{ and } G \text{ and } G)$$

$$P(2) = \frac{5!}{2!(5-2)!} \times \frac{10 \times 9 \times 90 \times 89 \times 88}{100 \times 99 \times 98 \times 97 \times 96}$$

$$P(2) = 10 \times 0.00702$$

$$P(2) = 0.0702$$

There are C^5_0 possible combinations of no bad and 5 good capacitors. The probability of exactly no bad capacitors in a sample of five is:

$$P(0) = C^5_0\, P(G \text{ and } B \text{ and } G \text{ and } G \text{ and } G)$$

$$P(0) = \frac{5!}{0!(5-0)!} \times \frac{90 \times 89 \times 88 \times 87 \times 86}{100 \times 99 \times 98 \times 97 \times 96}$$

$$P(0) = 1 \times 0.5838$$

$$P(0) = 0.5838$$

$P(1) = 0.3395$ from the calculation in Example 5-8

$$P(2 \text{ or less}) = P(2) + P(1) + P(0)$$

$$P(2 \text{ or less}) = 0.0702 + 0.3395 + 0.5838$$

Answer:

$$P(2 \text{ or less}) = 0.9935$$

The procedure for calculating the probability that a certain sample size will contain a specific number or less (or a specific number or more) bad parts when given or assuming the number of good and bad items in the population is as follows:

- Define the events in the probability space. These events will be that a sample contain no bad items to all bad (unless the population has fewer bad items than the sample size).
- Determine which events satisfy the question.
- The probability of each event is equal to the number of possible samples times the probability that one of those samples will be drawn.
- Add the event probabilities.

Example 5-10

Given:

One lot of 200 electric switches is to be sampled. The lot has 50 bad switches.

Find:

a. The probability that a sample of six will have 4 or more bad switches.
b. The probability that a sample of six will have 3 or less bad switches.

Solution:

► Define the events in the probability space.

A sample of six can only have:

exactly 6 bad switches

exactly 5 bad switches

exactly 4 bad switches

exactly 3 bad switches

exactly 2 bad switches

exactly 1 bad switches

exactly 0 bad switches

► Determine which events satisfy the question.

Four or more bad switches = 4, 5, or 6 bad switches.

$P(4\ or\ more) = P(4) + P(5) + P(6)$

► The probability of each event is equal to the number of possible sample times the probability that one of those samples will be drawn.

$$P(4) = C_4^6 P(B \text{ and } B \text{ and } B \text{ and } B \text{ and } G \text{ and } G)$$

$$P(4) = \frac{6!}{4!(6-4)!} \times \frac{50 \times 49 \times 48 \times 47 \times 150 \times 149}{200 \times 199 \times 198 \times 197 \times 196 \times 195}$$

$$P(4) = \frac{6 \times 5 \times 4 \times 3 \times 2 \times 1}{4 \times 3 \times 2 \times 1(2 \times 1)} \times 0.00208$$

$$P(4) = 15 \times 0.00208$$

$$P(4) = 0.0312$$

$$P(5) = C_5^6 P(B \text{ and } B \text{ and } B \text{ and } B \text{ and } B \text{ and } G)$$

$$P(5) = \frac{6!}{5!(6-5)!} \times \frac{50 \times 49 \times 48 \times 47 \times 46 \times 150}{200 \times 199 \times 198 \times 197 \times 196 \times 195}$$

$$P(5) = \frac{6 \times 5 \times 4 \times 3 \times 2 \times 1}{5 \times 4 \times 3 \times 2 \times 1(1)} \times 0.000643$$

$$P(5) = 6 \times 0.000643$$

$$P(5) = 0.00386$$

$$P(6) = C_6^6 P(B \text{ and } B \text{ and } B \text{ and } B \text{ and } B \text{ and } B)$$

$$P(6) = \frac{6!}{6!(6-6)!} \times \frac{50 \times 49 \times 48 \times 47 \times 46 \times 45}{200 \times 199 \times 198 \times 197 \times 196 \times 195}$$

$$P(6) = 1 \times 0.000193$$

$$P(6) = 0.000193$$

► Add the event probabilities.

$$P(4 \text{ or more}) = P(4) + P(5) + P(6)$$

$$P(4 \text{ or more}) = 0.0312 + 0.00386 + 0.000193$$

Answer:

$$P(4\ or\ more) = 0.0353$$

Since the probability space is equal to one, the probability of a sample of six containing 3 or less bad switches can be calculated easily.

$$P(4\ or\ more) + P(3\ or\ less) = 1$$

$$P(3\ or\ less) = 1 - P(4\ or\ more)$$

$$P(3\ or\ less) = 1 - 0.0353$$

Answer:

$$P(3\ or\ less) = .9647$$

This can be verified the long way:

$$P(3) = C_3^6\, P(B\ and\ B\ and\ B\ and\ G\ and\ G\ and\ G)$$

$$P(3) = \frac{6!}{3!(6-3)!} \times \frac{50\times 49\times 48\times 150\times 149\times 148}{200\times 199\times 198\times 197\times 196\times 195} = 0.131$$

$$P(2) = C_2^6\, P(B\ and\ B\ and\ G\ and\ G\ and\ G\ and\ G)$$

$$P(2) = \frac{6!}{2!(6-2)!} \times \frac{50\times 49\times 150\times 149\times 148\times 147}{200\times 199\times 198\times 197\times 196\times 195} = 0.301$$

$$P(1) = C_1^6 P(B\ and\ B\ and\ B\ and\ B\ and\ B\ and\ B\ )$$

$$P(1) = \frac{6!}{1!\ (6-1)!} \times \frac{50\times 150\times 149\times 148\times 147\times 146}{200\times 199\times 198\times 197\times 196\times 195} = 0.359$$

$$P(0) = C_0^6 P(B\ and\ B\ and\ B\ and\ B\ and\ B\ and\ B)$$

$$P(0) = \frac{6!}{0!\ (6-0)!} \times \frac{150\times 149\times 148\times 147\times 146\ \times 145}{200\times 199\times 198\times 197\times 196\times 195} = 0.173$$

The sum of the probabilities of each event in the probability space is 1.

P(exactly 6 bad switches)	=	0.000193
P(exactly 5 bad switches)	=	0.00386
P(exactly 4 bad switches)	=	0.0312

P(exactly 3 bad switches)	=	0.131
P(exactly 2 bad switches)	=	0.301
P(exactly 1 bad switch)	=	0.359
P(exactly 0 bad switches)	=	0.173
P(S)	=	1.00

Chapter Review

Keywords

Probability The relative frequency of a certain result or event in the long run.

Probability space All possible events that can occur. P(S) = 1

Independent event An event whose probability is not changed by the occurrence or non-occurrence of a prior event.

Mutually exclusive events Events that cannot occur simultaneously because getting one precludes getting the other.

Formulas

$$P(A) = \frac{\text{number of events that satisfy A}}{\text{number of events in the probability space}}$$

$$P(A \text{ and } B) = P(A) \times P(B|A)$$

$$P(A \text{ or } B) = P(A) + P(B) - P(AB)$$

$$C_r^n = \frac{n!}{r!\,(n-r)!}$$

Problems

5-1 What is the probability of getting a six on a single die?

5-2 What is the probability of getting two sixes on a pair of dice?

5-3 In a hand of 5 cards from a standard 52-card deck, what is the probability of getting four aces?

5-4 How many different hands of 13 cards are possible with a standard 52-card deck?

5-5 In a hand of 13 cards from a standard 52-card deck, what is the probability of getting two aces?

5-6 In a hand of 5 cards from a standard 52-card deck, what is the probability of getting at least two aces?

5-7 One lot of 100 items is received. You select 5 at random to inspect. If 20 of the items in the lot are bad, what is the probability that your sample will have:

a. no bad items?

b. two or less bad items?

c. three or more bad items?

d. exactly one bad item?

5-8 A container holds 50 balls: 10 are red, 5 are blue, and the remainder are transparent. If you do not replace selected balls, what is the probability that the selection of 4 balls produces:

a. only 1 red ball?

b. 4 red balls?

c. no blue balls?

5-9 A container holds 10 white and 80 black balls. Selected balls are not returned. What is the probability that a sample of 6 will contain:

a. exactly 3 white balls?

b. no white balls?

c. no black balls?

d. 3 or more black balls?

5-10 What would the probabilities be in problem 5 - 9 if each ball were returned to the container before the next selection?

6 Control Chart Sample Size

This chapter covers the determination of sample sizes based on the confidence that a control chart will detect a process change in a given number of samples.

Objectives

After completing this chapter you will be able to:

- calculate the probability that a single sample will detect a given shift of the process mean.
- calculate the sample size and number of samples that may be taken before a given shift of the process mean will be detected at a given level of confidence.
- develop a control chart to provide a given level of confidence that a specified process mean shift will be detected.

Process Mean Shift

In Chapter 4 we selected the control chart limits to be the average mean plus and minus three standard deviations of the sample means. The average sample mean is or is very close to the process mean. 99.7% of the sample means should fall between these limits as long as the process remains unchanged. This can be stated as "there is a 99.7% probability that if sample mean falls outside the control chart limits, the process is out of control." This probability of an event is called a *confidence level.*

Figure 6.1 shows a sample mean distribution on a control chart where the average sample mean is at the centerline. 99.7% of the sample means should fall between the upper and lower control limits. There is a 99.7% level of confidence that a sample average falling outside the limits means a process change. There is *not*, however, a 99.7% level of confidence that a sample average falling within the limits means that the process has *not* changed.

Figure 6.2 shows a sample mean distribution on a control chart where the average sample mean is above the centerline. Let's say that the process mean has shifted up due to a minor process change. Most of the sample mean distribution is still between the control limits. The probability of detecting this shift is the probability that a sample mean from the distribution will fall outside the control limits. The percentage of sample means outside the control limits is the shaded area of Figure 6.2.

We will call the shifted population mean $\bar{X}'s$. This area above the upper control limit can be calculated as the area under the normal curve between the upper control limit and the $\bar{X}'s + 3\,\sigma_{\bar{X}}$.

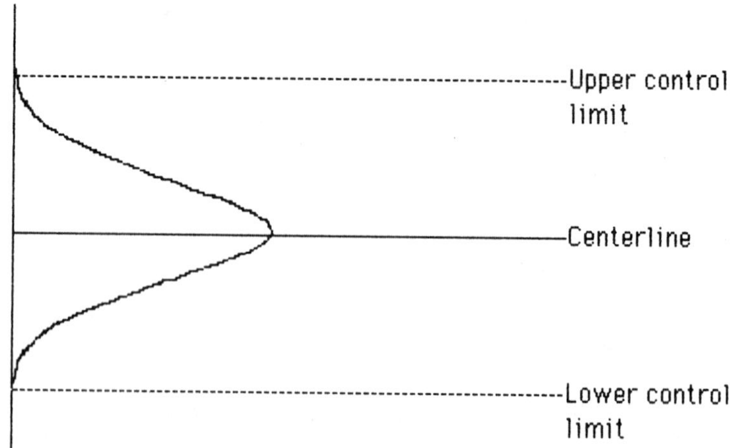

Figure 6.1 Control Chart with Sample Distribution

If this area is equal to 0.10, we have a 10% probability or confidence level of detecting the mean shift in the first sample after the shift occurred and a 90% chance of not detecting it. The probability of *not* detecting the shift in two samples is:

$$P(\textit{not detecting in } 2) = P(\textit{not detecting in 1st and not detecting in 2nd})$$

$$P(\textit{not detecting in } 2) = P(\textit{not detecting in 1st}) \times P(\textit{not detecting in 2nd})$$

$$P(\textit{not detecting in } 2) = 0.9 \times 0.9 = 0.81$$

$$P(\textit{ detecting in } 2) = 1 - P(\textit{ not detecting in } 2)$$

The probability of detecting the shift after two samples is 0.19. It would take seven samples to raise the probability of detection to 52%.

Figure 6.3 shows a sample mean distribution with a large area above the upper control limit. The probability of detecting this shift on the first sample is about 90%. The probability of not detecting it in the first sample after the shift is 0.10:

$$P(\textit{not detecting in } 2) = P(\textit{ not detecting in 1st}) \times P(\textit{not detecting in 2nd})$$

$$P(\textit{ not detecting in } 2) = 0.1 \times 0.1 = 0.01$$

$$P(\textit{detecting in } 2) = 1 - P(\textit{not detecting in } 2)$$

The probability of detecting the shift after two samples is 0.99.

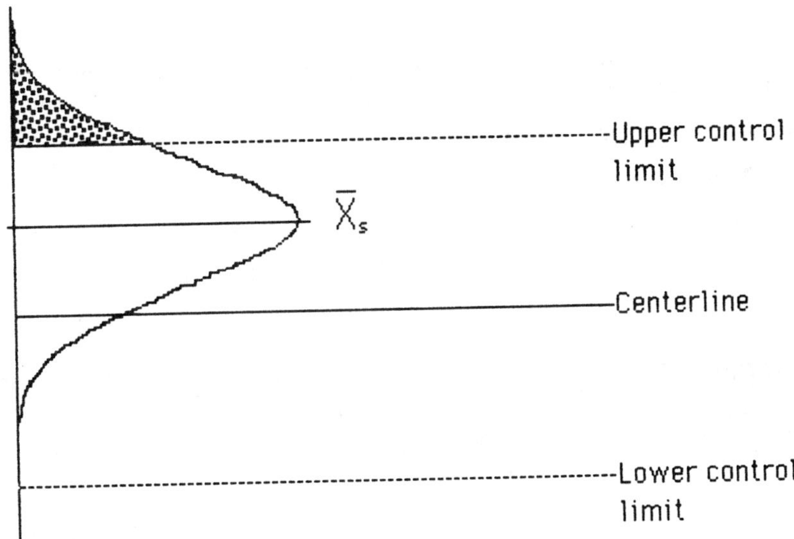

Figure 6.2 Control Chart with Shifted Sample Distribution

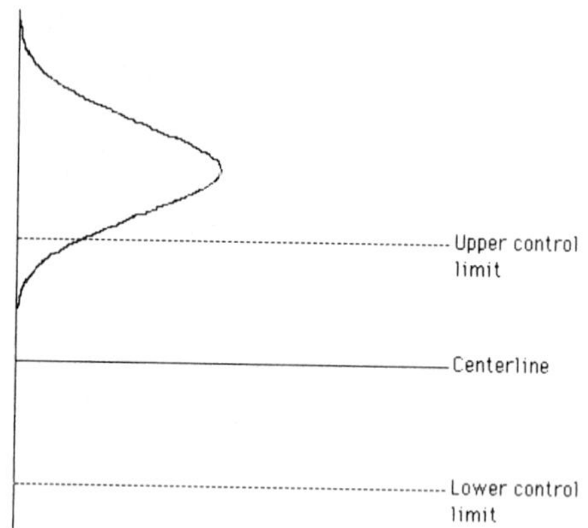

Figure 6.3 Control Chart with Shifted Sample Distribution

We would say that the sampling plan has a 90% confidence level of detecting the mean shift in the first sample and a 99% confidence level of detecting it in two samples. The maximum number of parts that are likely to be produced before detection of the shift is that number produced between the mean shift and the second sample.

How much a mean shift we are concerned about depends on how close the process limits are to specifications and the cost of scrap and rework. If the process mean shift shown in Figure 6.3 does not result in production outside the specifications, the process can be brought back in control and all items will be acceptable.

Improving the Probability of Detecting a Mean Shift

The main production goal is to limit the number of bad items made before a process change is detected. There are three ways to improve the probability of detecting a mean shift.

1. Increase the sample size.
2. Increase the sample frequency.
3. Use two sigma limits on the control chart.

Sample Size

The area of the shifted sample mean distribution that is above a control chart limit can be calculated as the area under the normal curve between the upper control limit and the $\bar{X}'s + 3\sigma_{\bar{X}}$.

The area of the shifted sample mean distribution that is below a control chart limit can be calculated as the area under the normal curve between the lower control limit and the $\bar{X}'s - 3\sigma_{\bar{X}}$. This area can be increased for a given mean shift by moving the control chart limits closer to the control chart centerline.

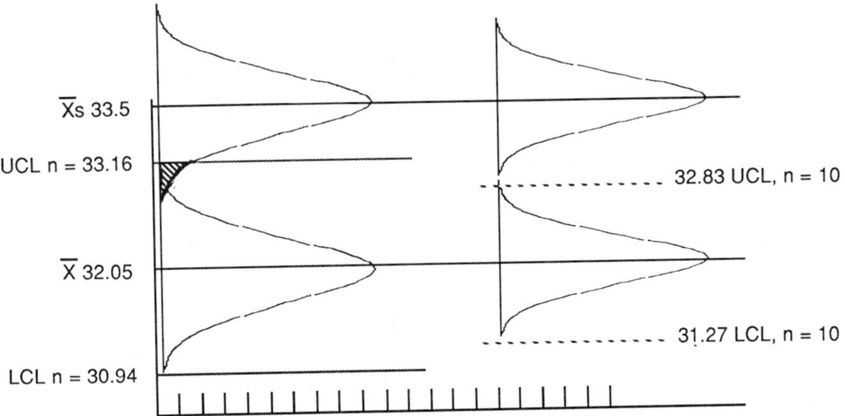

Figure 6.4 Comparison of Control Chart Limits with n = 5 and n = 10

The control chart limits for an $\bar{X}$ chart are calculated from sample range data by the equations:

$$UCL_{\bar{X}} = \bar{\bar{X}} + A_2 \bar{R}$$

$$LCL_{\bar{X}} = \bar{\bar{X}} - A_2 \bar{R}$$

A review of Appendix A, Table B, shows that the multiplier gets smaller as the sample size gets bigger. As a larger sample size brings the control limits closer to the centerline, it also decreases the value of $\sigma_{\bar{X}}$ since $\sigma_{\bar{X}} = \sigma' / \sqrt{n}$.

The lower left portion of Figure 6.4 shows the distribution of sample means from the $\bar{X}$ CHART in Figure 6.3. The upper left portion shows this distribution with the mean shifted to 33.5. The probability of detecting this shift in a single sample is equal to the area of the distribution above the upper control limit, 33.16. This area can be calculated from Table A.

For the area above the upper control limit:

$$Z = \frac{\bar{X}'s - UCL}{\sigma_{\bar{X}}}$$

For the area below the lower control limit:

$$Z = \frac{LCL - \bar{X}'s}{\sigma_{\bar{X}}}$$

Procedure:

The procedure to calculate the probability of detecting a population mean shift to a given value is:

- Calculate the sample mean standard deviation.
- Calculate the control chart limits.
- Calculate $\sigma_{\overline{X}}$.
- Calculate Z, the number of standard deviations between the nearest control limit and the shifted mean value.
- Determine the area under the normal curve from Table A.

Example 6-1

Given:

$n = 5$

$\overline{R} = 1.919$

$\overline{\overline{X}} = 32.05$

$\overline{X}'s = 33.5$

Find:

The probability of detecting a mean shift to 33.5 in the first sample taken after the shift.

Solution:

- Calculate the sample mean standard deviation.

The value of $\sigma_{\overline{X}}$ can be calculated from the sample means, from the control chart limits, or from the estimated population standard deviation.

- Calculate control chart limits.

$$UCL_{\overline{X}} = \overline{\overline{X}} + A_2\overline{R} = 32.05 + 0.577 \times 1.919 = 33.16$$

$$UCL_{\overline{X}} = \overline{\overline{X}} + A_2\overline{R} = 32.05 + 0.577 \times 1.919 = 30.94$$

► Calculate $\sigma_{\bar{X}}$

From the control chart limits:

$$\sigma_{\bar{X}} = \frac{UCL - LCL}{6} = \frac{33.16 - 30.94}{6} = 0.37$$

From the estimated population data we can calculate $\sigma_{\bar{X}}$ from the formula:

$\sigma_{\bar{X}} = \sigma' / \sqrt{n}$ where $\sigma' = \bar{R} / d_2$.

The $\bar{R}$ for this data is 1.919. The sample size is five. From Table B, $d_2 =$ 2.326.

$\sigma' = \bar{R} / d_2 = 1.919 / 2.326$

$\sigma' = 0.825$

$\sigma'_{\bar{X}} = \sigma' / \sqrt{n} = 0.825 / \sqrt{5}$

$\sigma_{\bar{X}} = 0.369$

► Calculate Z, the number of standard deviations between the nearest control limit and the shifted mean value.

$$Z = \frac{\bar{X}'s - UCL}{\sigma_{\bar{X}}}$$

$$Z = \frac{33.5 - 33.16}{0.369}$$

$Z = 0.921$

► Determine the area under the normal curve from Table A. From Table A, the area = 0.82.

Answer:

There is an 82% probability that the mean shift will be detected in one sample.

The effect of changing sample size is shown on the right side of Figure 6.4. Since the range data used for the original limits was based on samples of five measurements, calculation of proposed limits for a different sample size must be based on estimated population σ' and $\bar{\bar{X}}$.

Using the sample data $\sigma' = 0.825$ from Example 6-1, the limits for samples of 10 were calculated as:

$$UCL_{\bar{X}} = \bar{\bar{X}} + A\sigma' = 32.05 + .949 \times 0.825 \qquad UCL_{\bar{X}} = 32.83$$

$$LCL_{\bar{X}} = \bar{\bar{X}} - A\sigma' = 32.05 - .949 \times 0.825 \qquad LCL_{\bar{X}} = 31.27$$

The probability of detecting a mean shift to 33.5 then becomes:

$$\sigma_{\bar{X}} = \frac{UCL - LCL}{6} = \frac{32.83 - 31.27}{6} = 0.260$$

$$Z = \frac{\bar{X}'s - UCL}{\sigma_{\bar{X}}} = \frac{33.5 - 32.83}{0.260} = 2.5$$

From Table A the area = 0.995. There is an 99.5% probability that the mean shift will be detected in one sample.

An equation to calculate a sample size for a given probability that the mean shift will be detected in one sample can be derived from this formula. The Z value for a given probability can be taken from Table A. Then, from the equation for the upper control limit:

$$Z = \frac{\bar{X}'s - UCL}{\sigma_{\bar{X}}} = \frac{\bar{X}'s - (\bar{\bar{X}} + 3\sigma_{\bar{X}})}{\sigma_{\bar{X}}}$$

$$Z\sigma_{\bar{X}} = \bar{X}'s - (\bar{\bar{X}} + 3\sigma\bar{X})$$

$$Z\sigma_{\bar{X}} + 3\sigma_{\bar{X}} = \bar{X}'s - \bar{\bar{X}}$$

$$\sigma_{\bar{X}}(Z + 3) = \bar{X}'s - \bar{\bar{X}}$$

$$\sigma_{\bar{X}} = \frac{\bar{X}'s - \bar{\bar{X}}}{Z + 3} = \frac{\sigma'}{\sqrt{n}}$$

$$\sqrt{n} = \frac{\sigma'(Z + 3)}{\bar{X}'s - \bar{\bar{X}}}$$

Sample size for the upper control limit is:

$$n = \left[\frac{\sigma' (Z + 3)}{\bar{X}'s - \bar{\bar{X}}} \right]^2$$

For the lower control limit the formula becomes:

$$n = \left[\frac{\sigma' (Z + 3)}{\bar{\bar{X}} - \bar{X}'s} \right]^2$$

Sample Frequency

Sample size and frequency are related. A larger sample size increases the probability that a mean shift will be detected on the first sample taken after the shift occurs. For a given sample size, the probability of escaping detection goes down with each sample taken. The decision about whether to use large samples or more frequent smaller ones is an economic one. Inspection is not free. An understanding of the stability of the process is also necessary to make this decision. Processes with a history of sudden changes justify closer inspection than more stable processes. Control chart histories of prior operations should provide the necessary information.

Procedure:

The procedure to calculate the probability of detecting a population mean shift in a given number of samples is:

- ► Calculate the probability of detection in a single sample.
- ► Calculate the probability of not detecting a shift in a single sample.
- ► Calculate the probability of detection in N samples is 1 - (the probability of detection in a single sample raised to the Nth power).

Example 6-2

Given:

The results of Example 6-1.

Find:

The probability of detecting a population mean shift to 33.5 in at least two samples.

Solution:

► Calculate the probability of detection in a single sample.

In Example 6-1 we calculated that there is a 0.82 probability that a mean shift from 32.05 to 33.5 will be detected on the first sample taken after the shift.

► Calculate the probability of not detecting a shift in a single sample.

$P(\text{not detect in 1sample}) = 1 - P(\text{detect in 1sample})$

$P(\text{not detect in 1sample}) = 1 - 0.82$

$P(\text{not detect in 1sample}) = 0.18$

► Calculate the probability of detection in N samples is 1 – (the probability of detection in a single sample raised to the Nth power).

$P(\text{detect in } N \text{ samples}) = 1 - P(\text{not detect on 1st})^N$

$P(\text{detect in 2 samples}) = 1 - (0.18)^2$

Answer:

$P(\text{detect in 2 samples}) = 0.97$

Changing the probability of detection in the first sample to 90% would change the probability of detecting the shift by the second sample to:

$$P(\text{detect in 2}) = 1 - P(\text{not detect on 1st})^2$$

$$= 1 - (1 - .90)^2$$

$$P(\text{detect in 2}) = 0.99$$

If we decide that we can accept a 90% probability of detecting the shift in one sample with almost certainty of detecting it by the second sample, we can calculate the sample size required.

Procedure:

The procedure to calculate the sample size required to provide a given level of confidence that a population mean shift to a certain value will be detected in the first sample after the shift is:

- Estimate the population standard deviation.
- Get from Table A the Z value for the area under the normal curve equal to the desired confidence level.
- Calculate n.

Example 6-3

Given:

The following data was calculated from the first 200 items produced.

$n = 5$

$\bar{R} = 1.919$

$\bar{\bar{X}} = 32.05$

$\bar{X}'s = 33.5$

Find:

The sample size required to provide a 90% confidence level of detecting a mean shift to 33.5 in the first sample taken after the shift.

Solution:

► Estimate the population standard deviation.

The population standard deviation can be estimated as being equal to the standard deviation of all the measurements taken, or it can be calculated from the sample group average sample range or average sample standard deviation as discussed in Chapter 3.

From the given data:

$\sigma' = \bar{R}/d_2$

From Table B, d_2 for n = 5 is 2.326

$\sigma' = \bar{R}/d_2 = 1.919/2.326$

$\sigma' = 0.825$

► Get from Table A the Z value for the area under the normal curve equal to the desired confidence level.

► Get from Table A the Z value for .90 is 1.29.

► Calculate n.

$$n = \left[\frac{\sigma'(Z+3)}{\bar{X}'s - \bar{\bar{X}}}\right]^2 = \left[\frac{0.825\ (1.29+3)}{33.5 - 32.05}\right]^2$$

Answer:

$n = 6$

From these calculations, we now have a choice of sampling plans with the same total number of items being inspected. We can inspect less frequently with a sample size of ten and a 99.5% level of confidence of detecting a mean shift on the first sample. We can inspect more frequently with a sample size of six and a 90% level of confidence of detecting the mean shift on the first sample and a 99% level of confidence of detecting it by the second sample.

Changing Control Chart Limits

Normal control chart limits are three standard deviations from the centerline. This reduces the probability, to 0.03%, of assuming a process has changed when it in fact has *not* because 99.73% of the sample means fall between the $3\sigma_{\bar{X}}$ limits. Changing the limits to two standard deviations would increase this probability to 4.54% because 95.46% of the sample means fall between $\bar{\bar{X}} \pm 2\sigma_{\bar{X}}$. If it would be expensive for the process to go out of control without detection, and it would also be expensive to increase sample sizes or frequency but it would not be expensive to check out the increased number of false alarms, the 2σ limits are to be considered. To calculate the 2σ limits, we multiply the control chart limit multipliers in Table B by 2/3.

Example 6-4

Given:

$n = 5$

$\bar{R} = 1.919$

$\bar{\bar{X}} = 32.05$

$\bar{X}'s = 33.5$

Find:

a. The 2σ limits for an $\bar{X}$ chart.

b. The probability of detecting a mean shift to 33.5 in the first sample taken after the shift.

Solution:

2σ limits are calculated:

$$CL_{\bar{X}} = \bar{\bar{X}} + \frac{2A_2\bar{R}}{3} = 32.05 + \frac{2 \times .577 \times 1.919}{3} = 32.79$$

$$LCL_{\bar{X}} = \bar{\bar{X}} - \frac{2A_2\bar{R}}{3} = 32.05 - \frac{2 \times .577 \times 1.919}{3} = 31.31$$

Answer a:

$$UCL_{\bar{X}} = 32.79$$

$$LCL_{\bar{X}} = 31.31$$

The probability of detecting a mean shift to 33.5 on a single sample becomes:

$$Z = \frac{\bar{X}'s - UCL}{\sigma_{\bar{X}}} = \frac{33.5 - 32.79}{0.369} = 1.92$$

Answer b:

There is a 97% probability of detecting the mean shift in a single sample of five.

This is very close to the 99.5% probability provided by increasing the sample size to ten. This cuts the sampling cost in half but increases the false alarm rate from 0.03% to 4.54%.

Control Charts With Given Confidence Levels

For most manufacturing processes, the method of manufacture selected can produce items well within the specifications. The control charts are used to detect process changes with a reasonable assurance that changes will be detected before many unuseable items are produced. There is historical knowledge of the reliability of the process and a routine procedure of taking samples on a regular basis of time (e. g., every hour) or production (e.g., every 500 pieces).

Procedure:

The procedure to develop a control chart in this situation is:

- Collect samples totaling about 150 items.
- Calculate the process capability.
- Compare the process capability with the specifications and determine which mean shift would most likely cause scrap or rework expenses.
- Calculate the sample size that would give a reasonable confidence level of detecting the mean shift in a single sample.
- Develop the control charts for that sample size with four as a minimum, unless sampling is very expensive.
- Calculate the probability of detecting the population mean shift with the sample size and control limits selected.

Example 6-5

Given:

A machine is used to manufacture shafts. The specifications for an outside diameter are 1.500 ± 0.002 inch. The process has a history of normally turning 1500 shafts similar to this before significant variations occur. Oversized shafts can be reworked but undersized shafts will be scrap. It has been decided that the control charts should provide a 90% confidence level that undersized shafts will be detected on the first sample. Thirty samples of five measurements have been taken with the following results:

$\overline{\overline{X}} = 1.500$

$\overline{R} = 0.001$

Find:

The sample size and control chart limits.

Solution:

► Collect samples totaling about 150 items.

► Calculate the process capability.

The process capability is $\overline{\overline{X}} \pm 3\,\sigma'$.

$\sigma' = \overline{R} \,/\, d_2 = 0.001/2.326$

$\sigma' = 0.00043$

The upper process limit $= \overline{\overline{X}} + 3\sigma' = 1.500 + 3 \times 0.00043$
$= 1.5013$

The lower process limit $= \overline{\overline{X}} - 3\sigma' = 1.500 - 3 \times 0.00043$
$= 1.4987$

► Compare the process capability with the specifications and determine which mean shift would most likely cause scrap or rework expenses.

The lower process limit is 1.4987. Scrap is not produced until this limit moves to the lower specification, 1.498. The mean shift to be detected is:

$\overline{X}'s = \textit{Lower spec.} + 3\,\sigma'$

$\bar{X}'_s = 1.498 + 3 \times 0.00043$

$\bar{X}'_s = 1.4993$

► Calculate the sample size that would give a reasonable confidence level of detecting the mean shift in a single sample.

From Table A, the Z value for .90 is 1.29.

$$n = \left[\frac{\sigma'(Z+3)}{\bar{\bar{X}} - \bar{X}'_s}\right]^2 = \left[\frac{0.00043(1.29+3)}{1.500 - 1.4993}\right]^2$$

$n = 7$

► Develop the control charts for that sample size with four as a minimum unless sampling is very expensive.

Since the given $\bar{R}$ was calculated for samples of five, we must use σ' in calculating the control chart limits.

$Centerline = \bar{\bar{X}} = 1.5000$

$UCL_{\bar{X}} = \bar{\bar{X}} + A\sigma' = 1.500 + 1.134 \times 0.00043$

$UCL_{\bar{X}} = 1.5005$

$LCL_{\bar{X}} = \bar{\bar{X}} - A\sigma' = 1.500 - 1.134 \times 0.00043$

$LCL_{\bar{X}} = 1.4995$

$Centerline = \bar{R} = 0.001$

$UCL_R = D_2\sigma' = 5.203 \times 0.00043$

$UCL_R = 0.00224$

$LCL_R = D_1\sigma' = 0.205 \times 0.00043$

$LCL_R = 0$

Answer:

Sample size is 7. Control chart limits are:

$\bar{X}$ Chart	**R Chart**
Centerline = 1.500	Centerline = 0.001
$UCL_{\bar{X}} = 1.5005$	$UCL_R = 0.00224$
$LCL_{\bar{X}} = 1.4995$	$LCL_R = 0$

► Calculate the probability of detecting the population mean shift with the sample size and control limits selected.

Since the sample size has to be an integer, the actual performance of the chart may vary slightly from the initial confidence level. It is a good idea to check the result.

For the area below the lower control limit:

$$Z = \frac{LCL - \bar{X}'s}{\sigma_{\bar{X}}}$$

$$\sigma_{\bar{X}} = \frac{UCL - LCL}{6} = \frac{1.5005 - 1.4995}{6} = 0.000167$$

$$Z = \frac{1.4995 - 1.4993}{0.000167} = 1.12$$

From Table A, the probability of detecting the shift on the first sample is 88%.

Chapter Review

Keywords

Confidence level The probability that a certain event will occur.

Process mean shift The change of the population mean to a new value.

Probability of detection The probability that a plotted sample value will be outside the control chart limits.

Formulas

The area above the upper control limit:

$$Z = \frac{\bar{X}'s - UCL}{\sigma_{\bar{X}}}$$

The area below the lower control limit:

$$Z = \frac{LCL - \overline{X}'s}{\sigma_{\overline{X}}}$$

Sample size to provide given level of confidence of detecting mean shift:

Upper control limit:

$$n = \left[\frac{\sigma' (Z + 3)}{\overline{X}'s - \overline{\overline{X}}} \right]^2$$

Lower control limit:

$$n = \left[\frac{\sigma' (Z + 3)}{\overline{\overline{X}} - \overline{X}'s} \right]^2$$

$$P(\text{detect in } N \text{ samples}) = 1 - P(\text{not detect on 1st})^N$$

Problems:

6-1 Control chart data shows a product with $\overline{\overline{X}} = 2.000$ and $\overline{R} = 0.20$. The sample size is five. Calculate control chart limits for $\overline{X}$ and R charts. What is the probability that a mean shift to 2.08 will be detected on the first sample taken after the shift occurred?

6-2 Control chart data shows a product with $\overline{\overline{X}} = 2.50$ and $\overline{R} = 0.20$. The sample size is five. What is the probability that a mean shift to 2.38 will be detected on the first sample taken after the shift occurred?

6-3 Control chart data shows a product with $\overline{\overline{X}} = 6.00$ and $\overline{R} = 0.50$. The sample size is five. What is the probability that a mean shift to 6.8 will be detected on the first sample taken after the shift occurred?

6-4 Initial control chart data shows a product with $\overline{\overline{X}} = 5.00$ and $\overline{R} = 0.10$. The sample size is six. What is the probability that a mean shift to 5.15 will be detected on the first sample taken after the shift occurred? What is the probability that a mean shift to 5.15 will be detected on the first or second sample taken after the shift occurred?

6-5 Initial control chart data shows a product with $\overline{\overline{X}} = 3.100$ and $\overline{R} = 0.09$. The sample size is five. What sample size would be required for an 80% probability of detecting a mean shift to 3.15 in one sample?

6-6 Initial control chart data shows a product with $\bar{\bar{X}} = 3.00$ and $\bar{R} = 0.008$. The sample size is four.

a. Calculate the control chart limits.
b. Calculate the process capability.
c. Calculate the sample size required to provide an 85% confidence level that a mean shift to 3.006 wil be detected in one sample.

6-7 Initial control chart data shows a product with $\bar{\bar{X}} = 4.5$ and $\bar{R} = 0.08$. The sample size is five.

a. Calculate the control chart limits.
b. Calculate the process capability.
c. Calculate the sample size required to provide a 90% confidence level that a mean shift to 4.55 will be detected in one sample.
d. Check the actual confidence level for detecting the mean shift with the sample size selected in c.

6-8 A container-filling process must provide at least 32 ounces of product in each container. Initial population data shows that $\bar{\bar{X}} = 32.100$ and $\sigma' = 0.020$. Develop control charts that would give a 99% confidence level of a single sample detecting a population mean shift which would violate the 32-ounce minimum.

6-9 Initial population data shows that a dimension has $\bar{\bar{X}} = 6.000$ and $\sigma' = 0.001$. The specifications for this dimension are 6.000 ± 0.005 inch. Develop control charts that will provide a 90% level of confidence of a single sample detecting a mean shift that would result in violating the specifications.

6-10 Control charts have been developed for a process. $\bar{\bar{X}} = 9.00$ and $\bar{R} = 0.08$ with a sample size of five.

a. Calculate the probability of detecting a mean shift to 9.05 in one sample.
b. Calculate the probability of detecting a mean shift to 9.05 in one sample if the charts were modified to two sigma limits.

7 Probability Distribution

This chapter introduces the binomial and Poisson probability distributions. These distributions calculate the probability of certain numbers of defects in a sample. They will be the basis of control charts for attributes to be discussed in Chapter 8.

Objectives

After completing this chapter you will be able to:

- use the binomial distribution to calculate the probability of a certain number of defects in a sample.
- use the Poisson distribution to calculate the probability of a certain number of defects in a sample.

The Binomial Distribution

In Example 5-10, we had a population of 150 good switches and 50 bad switches. We calculated the probability that a sample of six items from this population would have various numbers of defects. We used the procedure that stated, "The probability of each event is equal to the number of possible samples times the probability that one of those samples will be drawn."

For the probability that a sample would contain exactly four bad items, our calculation was:

$$P(4) = C^6{}_4\, P(\ B \ and\ B\ and\ B\ and\ B\ and\ G\ and\ G\ )$$

$$P(4) = \frac{6!}{4!(6\text{-}4)!} \times \frac{50 \times 49 \times 48 \times 47 \times 150 \times 149}{200 \times 199 \times 198 \times 197 \times 196 \times 195}$$

$$P(4) = \frac{6 \times 5 \times 4 \times 3 \times 2 \times 1}{4 \times 3 \times 2 \times 1 \times (2 \times 1)} \times 0.00208$$

$$P(4) = 15 \times 0.00208$$

$$P(4) = 0.0312$$

If we had replaced each item after inspecting it and before drawing the next sample, the calculation would have been:

$$P(4) = \frac{6!}{4!(6\text{-}4)!} \times \frac{50 \times 50 \times 50 \times 50 \times 150 \times 150}{200 \times 200 \times 200 \times 200 \times 200 \times 200}$$

$$P(4) = 15 \times .00219$$

$$P(4) = 0.03296$$

The solution 0.03296 is almost six percent higher than the first solution.

If we had a larger population with the same proportion of bad items, say a population of 2000 and 500 defects, the first calculation becomes:

$$P(4) = C^6{}_4\, P(\ B \ and\ B\ and\ B\ and\ B\ and\ G\ and\ G\ )$$

$$P(4) = \frac{6!}{4!(6\text{-}4)!} \times \frac{500 \times 499 \times 498 \times 497 \times 1500 \times 1499}{2000 \times 1999 \times 1998 \times 1997 \times 1996 \times 1995}$$

$$P(4) = \frac{6 \times 5 \times 4 \times 3 \times 2 \times 1}{4 \times 3 \times 2 \times 1 \times (2 \times 1)} \times 0.0021859$$

$$P(4) = 0.03279$$

The solution with replacement remains the same because the proportion of bad items did not change: 50/200 is the same value as 500/2000. With this larger population, the error of assuming replacement is only one half of one percent. In a large population, the effect of diminishing the population by taking a sample is less significant that it is in a small population. Assuming replacement and ignoring the small error caused by diminishing the population makes calculating the probability easier.

Let's take a second look at the calculation with replacement:

$$P(4) = \frac{6!}{4!(6\text{-}4)!} \times \frac{\mathit{50}}{\mathit{200}} \times \frac{\mathit{50}}{\mathit{200}} \times \frac{\mathit{50}}{\mathit{200}} \times \frac{\mathit{50}}{\mathit{200}} \times \frac{150}{200} \times \frac{150}{200}$$

The four fractions in italics are the proportion of bad items. We'll call this fraction of bad or nonconforming items in the population p', p prime. In this case $p' = 0.25$. The remaining two fractions are the proportion of good items. We'll call this proportion $(1 - p')$. In some texts it is also called q'. In this case $(1 - p') = 0.75$.

The calculation now becomes:

$$P(4) = \frac{6!}{4!(6\text{-}4)!} \times p'^4(1-p')^2$$

The exponent, 4, is equal to r, the number of bad items in the sample of six. The exponent, 2, is equal to the number of good items which is the sample size, n, minus the number of bad items, r.

The general formula for the probability of exactly r bad items in a sample of n is the binomial probability formula:

$$\frac{n!}{r!(n\text{-}r)!} p'^r(1-p')^{(n-r)}$$

Example 7-1

Given:

One lot of 3000 items with 20 percent defective items is to be inspected with one sample of 20.

Find:

The binomial probability that the sample will contain 5 or less defective items.

Solution:

$$P(5 \text{ or less}) = P(0, 1, 2, 3, 4, \text{or } 5)$$

$$P(5 \text{ or less}) = P(0) + P(1) + P(2) + P(3) + P(4) + P(5)$$

$$P(exactly\,r) = \frac{n!}{r!\,(n\text{-}r)!} p'r(1 - p')^{(n\text{-}r)}$$

$$P(0) = \frac{20!}{0!\ (20-0)!} 0.2^0(1-0.2)^{(20\text{-}0)}$$

$$P(0) = 1 \times 1(0.8)^{20}$$

$$P(0) = 0.011529$$

$$P(1) = \frac{20!}{1!\ (20-1)!} 0.2^1(1-0.2)^{(20\text{-}1)}$$

$$P(1) = 20 \times 0.2 \times 0.8^{19}$$

$$P(1) = 0.057646$$

$$P(2) = \frac{20!}{2!\,(20-2)!} 0.2^2(1-0.2)^{(20\text{-}2)}$$

$$P(2) = 190 \times 0.04 \times 0.8^{18}$$

$$P(2) = 0.136909$$

$$P(3) = \frac{20!}{3!\,(20-3)!} 0.2^3(1-0.2)^{(20\text{-}3)}$$

$$P(3) = 1140 \times 0.008 \times 0.8^{17}$$

$$P(3) = 0.205364$$

$$P(4) = \frac{20!}{4!\ (20-4)!} 0.2^4(1-0.2)^{(20\text{-}4)}$$

$$P(4) = 4845 \times 0.0016 \times 0.8^{16}$$

$$P(4) = 0.218199$$

$$P(5) = \frac{20!}{5!(20-5)!} 0.2^5 (1-0.2)^{(20-5)}$$

$$P(5) = 15504 \times 0.00032 \times 0.8^{15}$$

$$P(5) = 0.1746$$

$$P(5\ or\ less) = P(0) + P(1) + P(2) + P(3) + P(4) + P(5)$$

$$P(0) = 0.011529$$

$$P(1) = 0.057646$$

$$P(2) = 0.136909$$

$$P(3) = 0.205364$$

$$P(4) = 0.218199$$

$$P(5) = \underline{0.1746}$$

$$0.8042$$

Answer:

$$P(5\ or\ less) = 0.8042$$

The binomial distribution has a standard deviation which is:

$$\sigma'_{np} = \sqrt{np'(1-p')}$$

We will use it in Chapter 8 to develop control charts.

The shape of the distribution depends on the sample size and p′ as shown in Charts 7.1 through 7.3. The average number of defective items in a sample is np′. In Example 7-1 this is:

$$np' = 20 \times 0.2$$

$$np' = 4$$

The standard deviation is:

$$\sigma'_{np} = \sqrt{4(1-0.2)}$$

$$O'_{np} = 1.79$$

For that data in Example 7-1, Figure 7.1 shows the probabilities of exactly R defects as columns and the cumulative probability of R or less defects as a line. Note that the column chart is not symmetrical.

Figure 7.2 shows the distribution for the same sample size but with p′ = 0.6. The probability distribution is more symmetrical.

Figure 7.3 is the distribution of a sample of 100 with p′ = 0.02. This distribution is not symmetrical at all. The major cause of the lack of symmetry is that the number of defects is a positive integer. If the value of np′ is close to 1, there is a high probability of zero defects in the sample. The values of P(R) and P(R or less) are shown in Table 7.1.

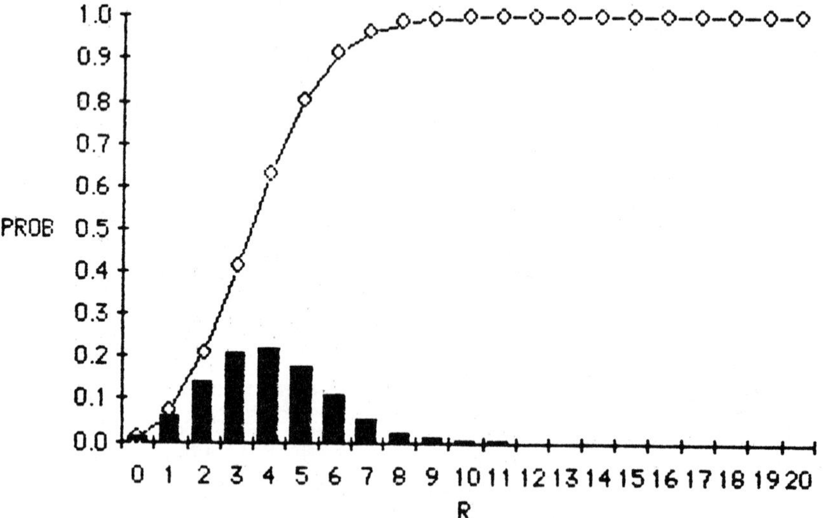

Figure 7.1 Probability of R defects

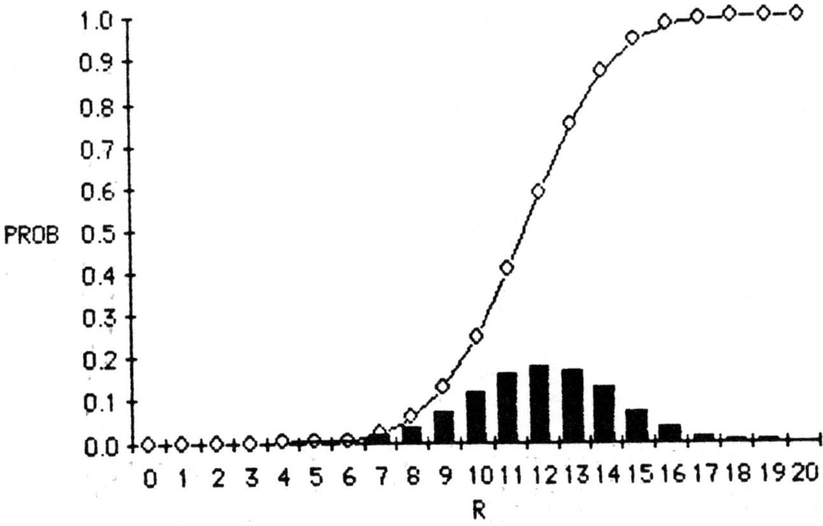

Figure 7.2 Probability of R defects

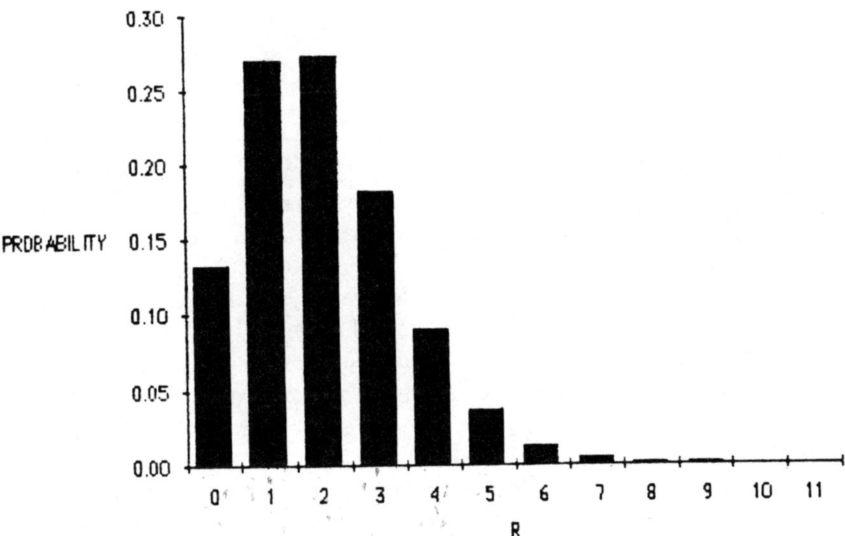

Figure 7.3 Probability of R defects

Table 7.1

$n = 100$	$p' = 0.02$	
R	**P(R)**	**P(R or less)**
0	.13262	.13262
1	.27065	.40327
2	.27341	.67669
3	.18228	.85896
4	.09021	.94917
5	.03535	.98452
6	.01142	.99594
7	.00313	.99907
8	.00074	.99981
9	.00015	.99997
10	.00003	.99999
11	0.00000	1.00000

The Poisson Distribution

In the binomial equation, p′ is the proportion or percentage of defective items. To calculate it we need to know the number of items in the population and how many of them are defective, or the rate at which defects occur in an ongoing process.

There are situations where there is no p′. If we are manufacturing photographic film and find six bubbles in a yard-long sample, we cannot calculate a p′. The yard of film could have an unlimited number of bubbles. Over time, however, we could develop a history of how many bubbles are found in the average yard.

We will call this expected or average number c′, c prime. The Poisson distribution will allow the calculation of the probability of finding a specific number of defects, r. The Poisson formula is:

$$P(r) = \frac{c'^{r} e^{-c'}}{r!}$$ where e = 2.718, the natural logarithm.

The standard deviation of the Poisson distribution is:

$$O'_c = \sqrt{c'}$$

Example 7-2

Given:

The average number of defects found in a yard of film is 6.0.

Find:

The probability that a yard will have 2 or less defects.

Solution:

$$P(2\ or\ less) = P(0, 1, 2)$$

$$P(2\ or\ less) = P(0) + P(1) + P(2)$$

$$P(r) = \frac{c'^{r} e^{-c'}}{r!}$$

$$P(0) = \frac{6^{0} \times 2.718^{-6}}{0!} = 0.00248$$

$$P(1) = \frac{6^{1} \times 2.718^{-6}}{1!} = 0.01488$$

$$P(2) = \frac{6^{2} \times 2.718^{-6}}{2!} = 0.04465$$

$$P(2\ or\ less) = P(0) + P(1) + P(2)$$

Answer:

$$P(2\ or\ less) = 0.062$$

Table 7.2 shows the probabilities of R for c′ = 6. These values are plotted in Figure 7.4.

Table 7.3 contains factorials for numbers from 1 to 49.

Chapter Review

Terms

p′ Percentage or proportion of a population that is defective.

c′ The average number of defects.

Formulas

Binomial

$$P(exactly\ r) = \frac{n!}{r!\,(n-r)!} p'^{r}(1-p')^{(n-r)}$$

$$\sigma'_{np} = \sqrt{np'(1-p')}$$

Table 7.2

R	*P(R)*
0	0.0024802947020988
1	0.014881768212593
2	0.044645304637778
3	0.089290609275557
4	0.13393591391334
5	0.160723096696
6	0.160723096696
7	0.13776265431086
8	0.10332199073314
9	0.068881327155431
10	0.041328796293257
11	0.022542979796322
12	0.011271489898161
13	0.0052022261068437
14	0.0022295254743616
15	0.00089181018974461
16	0.00033442882115422
17	0.00011803370158385

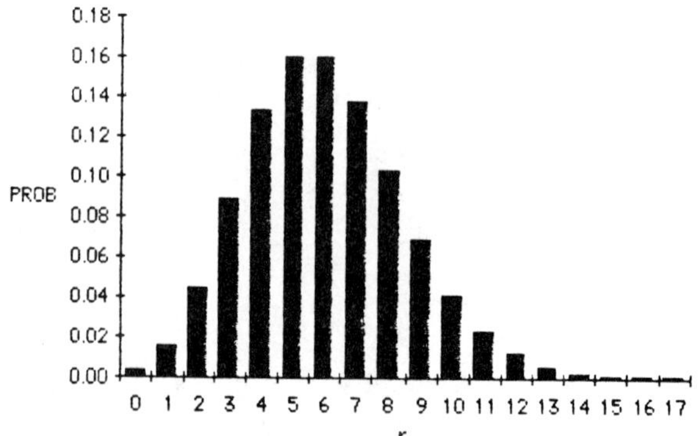

Figure 7.4 Poisson Probability

Table 7.3 Factorials for Numbers 1-49

n	$n!$
0	1
1	1
2	2
3	6
4	24
5	120
6	720
7	5040
8	40320
9	362880
10	3628800
11	39916800
12	479001600
13	6227020800
14	87178291200
15	1307674368000
16	20922789888000
17	355687428096000
18	6402373705728000
19	121645100408830000
20	2432902008176600000
21	5.1090942171709E+19
22	1.1240007277776E+21
23	2.5852016738885E+22
24	6.2044840173324E+23
25	1.5511210043331E+25
26	4.0329146112661E+26

Table 7.3 Continued

n	$n!$
27	1.0888869450418E+28
28	3.048883446117E+29
29	8.8417619937393E+30
30	2.6525285981218E+32
31	8.2228386541776E+33
32	2.6313083693368E+35
33	8.6833176188114E+36
34	2.9523279903959E+38
35	1.0333147966386E+40
36	3.719933267899E+41
37	1.3763753091226E+43
38	5.2302261746659E+44
39	2.0397882081197E+46
40	8.1591528324788E+47
41	3.3452526613163E+49
42	1.4050061177528E+51
43	6.041526306337E+52
44	2.6582715747883E+54
45	1.1962222086547E+56
46	5.5026221598116E+57
47	2.5862324151115E+59
48	1.2413915592535E+61
49	6.0828186403422E+62

Poisson

$$P(r) = \frac{c'^{r} e^{-c'}}{r!} \text{ where e} = 2.718\text{, the natural logarithm}$$

$$\sigma'_c = \sqrt{c'}$$

7-1 A sample of 20 is taken from a population with 30% defective items. What is the binomial probability that the sample will contain 3 or less defective items?

7-2 A sample of 15 is taken from a population with 10% defective items. What is the binomial probability that the sample will contain 3 or less defective items?

7-3 A lot to be inspected is assumed to have 10% defective items. A sample of 10 is taken and has 2 defective items. What is the binomial probability that a sample of 10 would have 2 or more defective items if the population has 10% defective items? Hint: P(2 or more) # 1 − P(1 or less).

7-4 A sample of 10 is taken from a population with 5% defective items. What is the probability that the sample will contain no defective items?

7-5 A sample of 10 is taken from a population with 8% defective items. What is the probability that the sample will contain no defective items?

7-6 The average number of seeds, (bubbles), in a certain size glass vase has been 3. What is the Poisson probability that a single vase will have 1 or less?

7-7 The average number of seeds, (bubbles), in a certain size glass vase has been 3. What is the Poisson probability that a single vase will have 5 or more?

7-8 The average July rainfall in a city has been three inches. What is the Poisson probability of getting no rain this July?

7-9 The average number of knots in a ball of yarn has been one. What is the probability that a ball from this population will have three or more knots?

7-10 An inspection plan states that a bolt of cloth will be accepted if it contains no more that one defect. What is the probability that a bolt from a population with an average of two defects will be accepted?

8 Control Charts for Attributes

This chapter covers the development and use of control charts for attributes. There are three major types of control charts used for attributes. The p chart is for the fraction of defective items in a sample. The np chart is for the number of defective items in a sample. The c chart is for the number of defects in an item. The u chart is for the number of defects in a sample.

Objectives

After completing this chapter you will be able to:

- calculate the limits for p charts.
- calculate the limits for np charts.
- calculate the limits for c charts.
- calculate the limits for u charts.
- calculate the sample size and/or sample frequency required to detect a population change with a given confidence level.

Fraction Defective Chart

The fraction defective is the number of defective items in a sample divided by the total number of items in the sample. It is always expressed as a decimal.

The fraction defective chart is used when the sample size varies. If we have a high percentage of good items, say 99%, the fraction defective is small, 0.01. In order to get any defectives in a sample from a high quality population, the sample size must be large. In many cases the sample size is all the daily production. In this situation the sample size will vary from day to day. The only statistical measure of quality would be the fraction rejected.

As we saw in Chapter 7, samples taken from a population with a fraction defective of p' will not always have the same number of defective items. The distribution of the sample fraction defective, called p, will closely follow the binomial distribution as long as the depletion of items from the population is not significant. As with other control charts, experience has shown that three-sigma control limits provide a reasonable balance between economy and reliability for most situations.

The standard deviation for the number of defective items in a sample (Chapter 7) is:

$$\sigma'_{np} = \sqrt{np'(1 - p')}$$

For the fraction defective, this becomes:

$$\sigma_p = \sqrt{\frac{p'(1 - p')}{n}}$$

As was the case with $\bar{X}'$ and $\bar{\bar{X}}$, the average value of fraction defective, $\bar{p}$, can be used in place of p' when the value of p' is not known or assumed. The equation then becomes:

$$\sigma_p = \sqrt{\frac{\bar{p}(1 - \bar{p})}{n}}$$

Using the value for the average fraction defective, the control limit formulas become:

$$UCL_p = \bar{p} + 3\sqrt{\frac{\bar{p}(1 - \bar{p})}{n}}$$

$$UCL_p = \bar{p} - 3\sqrt{\frac{\bar{p}(1 - \bar{p})}{n}}$$

There are three methods of accommodating the changes in sample size:

1. calculate chart limits for the average sample size and double-check the limits for points near the control limits;
2. calculate new limits for each sample;
3. plot limits for a range of sample sizes on the chart and record the sample size with the plotted p values.

Example 8-1

Given:

Production of a plastic container has an average fraction defective of 0.0097. Production data for a 20-day period is given in Table 8.1.

Find:

Plot the p chart for the data with the limits based on each day's production as the sample size.

Solution:

For all days, the value of $\bar{p}$ is 0.0097. The value of n changes with each day's production. The value of p = n / (No. of defectives). For day 1:

$$p = 26 / 2567$$

$$p = 0.0101$$

$$UCL_p = \bar{p} + 3\sqrt{\frac{\bar{p}(1-\bar{p})}{n}}$$

$$UCL_p = 0.0097 + 3\sqrt{\frac{0.0097(1-0.0097)}{2567}}$$

$$UCL_p = 0.0155$$

$$LCL_p = \bar{p} - 3\sqrt{\frac{\bar{p}(1-\bar{p})}{n}}$$

$$LCL_p = 0.0097 - 3\sqrt{\frac{0.0097(1-0.0097)}{2567}}$$

$$LCL_p = 0.0039$$

The values for this and the remaining days are calculated in the same manner and are shown in Table 8.2.

Table 8.1 Production Data for a 20-day Period

Day	*Items Inspected*	*No. Defective*
1	2567	26
2	2098	21
3	2876	27
4	1987	17
5	2134	19
6	2165	19
7	2343	26
8	2234	23
9	2189	22
10	2543	28
11	2453	25
12	2541	29
13	2179	20
14	2065	21
15	2477	23
16	2561	25
17	2073	17
18	2331	21
19	2454	24
20	2476	21

Table 8.2 Fraction Defective (p) Values and Limits

Day	*n*	*No. defective*	*p*	*UCL*	*LCL*
1	2567	26	0.0101	0.0155	0.0039
2	2098	21	0.0100	0.0161	0.0033
3	2876	27	0.0094	0.0152	0.0042
4	1987	17	0.0086	0.0163	0.0031
5	2134	19	0.0089	0.0161	0.0033
6	2165	19	0.0088	0.0160	0.0034
7	2343	26	0.0111	0.0158	0.0036
8	2234	23	0.0103	0.0159	0.0035
9	2189	22	0.0101	0.0160	0.0034
10	2543	28	0.0110	0.0155	0.0039
11	2453	25	0.0102	0.0156	0.0038
12	2541	29	0.0114	0.0155	0.0039
13	2179	20	0.0092	0.0160	0.0034
14	2065	21	0.0102	0.0162	0.0032
15	2477	23	0.0093	0.0156	0.0038
16	2561	25	0.0098	0.0155	0.0039
17	2073	17	0.0082	0.0162	0.0032
18	2331	21	0.0090	0.0158	0.0036
19	2454	24	0.0098	0.0156	0.0038
20	2476	21	0.0085	0.0156	0.0038

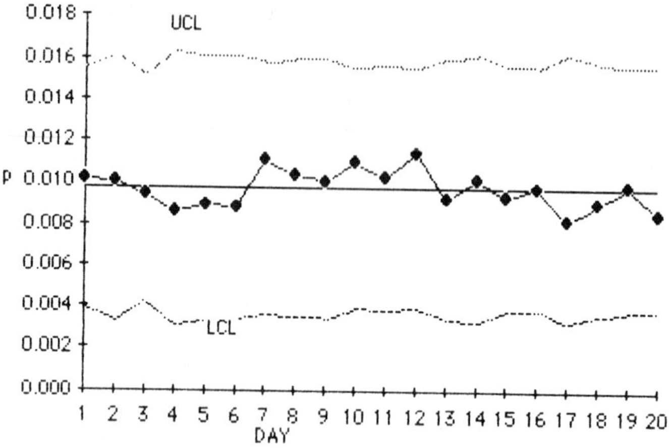

Figure 8.1 p Chart with Moving Control Limits

Answer:

Figure 8.1.

The average sample size for the twenty days is 2337. The limits for day 11 are approximately the value that would have been plotted if an average day's production was used to calculate the control limits instead of each day's production (Figure 8.1).

Example 8-2

Given:

Production of a casting has an average fraction defective of 0.1. Production data for a 15-day period is given in Table 8.3. Average daily production has been about 830 castings.

Find:

Plot the p chart for the data with the limits based on average daily production as the sample size.

Solution:

$$UCL_p = \bar{p} + 3\sqrt{\frac{\bar{p}(1-\bar{p})}{n}}$$

$$UCL_p = 0.1 + 3\sqrt{\frac{0.1(1-0.1)}{830}}$$

$$UCL_p = 0.1312$$

Table 8.3 Production Data for a 15-day Period

Day	*Number Inspected*	*Number Defective*
1	823	81
2	902	85
3	750	87
4	878	91
5	789	82
6	821	98
7	856	79
8	824	70
9	895	86
10	785	82
11	798	77
12	834	87
13	822	83
14	796	72
15	854	88

$$LCL_p = \bar{p} - 3\sqrt{\frac{\bar{p}(1-\bar{p})}{n}}$$

$$LCL_p = 0.1 - 3\sqrt{\frac{0.1(1-0.1)}{830}}$$

$$LCL_p = 0.0688$$

For day 1

p = No. Defectives/No. inspected
p = 81/823
p = 0.0984

The values of p for each day are given in Table 8.4.

Answer:

Figure 8.2.

Table 8.4 Fraction Defective (p) Values

DAY	*n*	*No. Defective*	*p*
1	823	81	0.0984
2	902	85	0.0942
3	750	87	0.1160
4	878	91	0.1036
5	789	82	0.1039
6	821	98	0.1194
7	856	79	0.0923
8	824	70	0.0850
9	895	86	0.0961
10	785	82	0.1045
11	798	77	0.0965
12	834	87	0.1043
13	822	83	0.1010
14	796	72	0.0905
15	854	88	0.1030

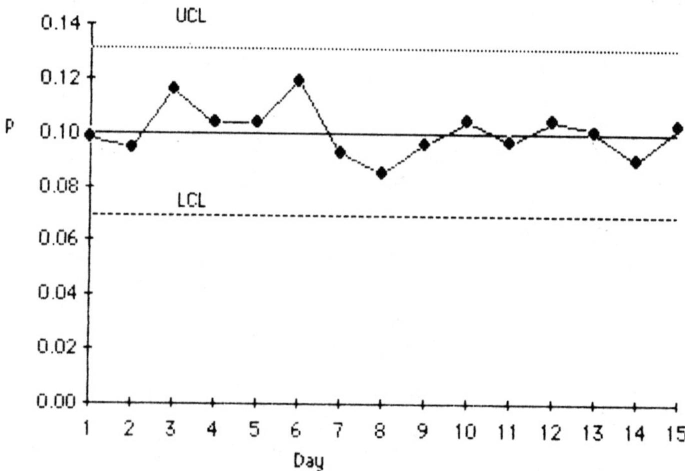

Figure 8.2 p Chart Using Average Sample Size

Number Defective Chart

The number defective, np, chart shows the number of defective items in samples rather than the fraction of defective items. It requires that the sample size remain constant. It has two benefits over the p chart: there is no calculation required of each sample result; it is easier for some people to understand.

The chart centerline is $n\bar{p}$, which is the average number of defects per sample. If the population has a known or assumed p', the centerline is np'.

The control limits are the three-sigma limits.

$$UCL_{np} = n\bar{p} + 3\sqrt{n\bar{p}(1 - \bar{p})}$$

$$LCL_{np} = n\bar{p} - 3\sqrt{n\bar{p}(1 - \bar{p})}$$

It should be noted that even though the control limits and centerline may be decimal numbers, the number of defects in a sample will always be an integer.

Example 8-3

Given:

Fifteen samples were taken from a population with a $\bar{p}$ of 0.0222. The sample size is 300. The sample data is in Table 8.5.

Find:

The np chart control limits. Plot the data on an np chart.

Table 8.5 Data from 15 Samples

Sample No.	*No. of Defects*	*Sample No.*	*No. of Defects*
1	1	9	4
2	12	10	9
3	7	11	10
4	9	12	7
5	3	13	8
6	2	14	4
7	8	15	9
8	7		

Solution:

$$\text{Centerline} = n\bar{p} = 300 \times 0.0222$$

$$\text{Centerline} = 6.66$$

$$UCL_{np} = n\bar{p} + 3\sqrt{n\bar{p}(1 - \bar{p})}$$

$$UCL_{np} = 6.66 + 3\sqrt{6.66(1 - 0.0222)}$$

$$UCL_{np} = 14.3$$

Since the values plotted on the np chart are integers, the control limits are plotted to only one decimal place so that a plotted value is definitely either above or below the limit.

$$LCL_{np} = n\bar{p} - 3\sqrt{n\bar{p}(1 - \bar{p})}$$

$$LCL_{np} = 6.66 - 3\sqrt{6.66(1 - 0.0222)}$$

$$LCL_{np} = -0.996$$

Since a sample cannot have less that zero defectives:

$$LCL_{np} = 0$$

Answer:

Figure 8.3.

Control Chart for Number of Defects, c Chart

The number of defects, c, chart is based on the Poisson distribution. It is a plot of the number of defects in items. The item may be a given length of steel bar, a welded tank, a bolt of cloth, and so on. For the control chart, the size of the item must be constant. If the chart is for the number of defects in a bolt of cloth, all the bolts must be of the same size.

The centerline of the c chart is c′ or c bar. The three-sigma control limits are:

$$UCL_c = c' + 3\sqrt{c'}$$

$$LCL_c = c' - 3\sqrt{c'}$$

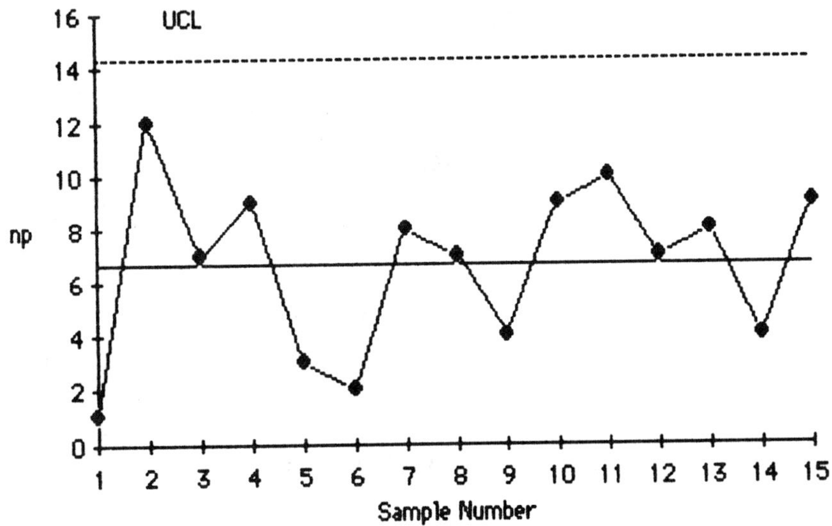

Figure 8.3 np Chart

Example 8-4

Given:

The average number of defects in a roll of film has been eight. Data for the last 15 rolls produced are in Table 8.6.

Find:

The c chart control limits and plot the data from Table 8.6.

Table 8.6 Data for 15 Rolls of Film

Roll No.	*No. of Defects*	*Roll No.*	*No. of Defects*
1	1	9	4
2	12	10	7
3	7	11	9
4	9	12	6
5	3	13	13
6	2	14	8
7	8	15	6
8	7		

Solution:

$$UCL_c = c' + 3\sqrt{c'}$$

$$UCL_c = 8 + 3\sqrt{8}$$

$$UCL_c = 16.5$$

$$LCL_c = c' - 3\sqrt{c'}$$

$$LCL_c = 8 - 3\sqrt{8}$$

Since this is a negative value,

$$LCL_c = 0$$

Answer:

Figure 8.4.

$$UCL_c = 16.5$$

$$LCL_c = 0$$

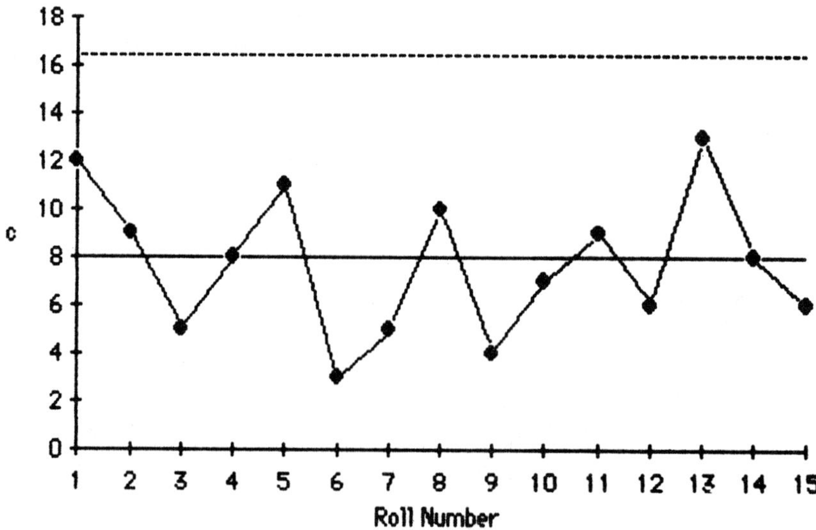

Figure 8.4 c Chart

Average Number of Defects Chart

The average number of defects, c, chart is for the number of defects in a single item. It can also be used for the number of defects in a fixed number of items. The symbol u is used to represent defects per unit. It is equal to c/n. The average value of u is $\bar{u}$.

$$\bar{u} = \frac{\Sigma c}{\Sigma n} = \frac{\textit{total defects}}{\textit{total of items inspected}}$$

The control limits are:

$$UCL = \bar{u} + \frac{3\sqrt{\bar{u}}}{\sqrt{n}}$$

$$LCL = \bar{u} - \frac{3\sqrt{\bar{u}}}{\sqrt{n}}$$

The c chart can be used for a constant sample size. The number of defects per 10 bolts of cloth can be plotted on c charts just as well as the number of defects per single roll. The essential factor for using c is that each sample have the same opportunity for defects.

The u chart is used in cases where the samples are of different size. Since the sample size is in the control chart limit formula, we have the same options we have with the p charts. If the sample size remains fairly constant, an average sample size can be used to calculate the u-chart limits. If the sample size varies significantly, each sample value must be plotted with its own limits. Example 8-5 contains data on tractor subassemblies where the daily production varies from one to six units.

Example 8-5

Given:

A fabrication shop makes tractor subassemblies. The average number of defects per tractor has been 4.14. Data for 10 day's production is in Table 8.7.

Find:

Develop a u chart and plot the data.

Solution:

Since this is not the initial u chart, the historical average of $\bar{u} = 4.14$ will be used to calculate the limits.

For day 1:

$$u = c/n = 29/4$$

$$u = 7.25$$

Table 8.7 Data for 10 Days' Production

DAY	*Total Defects* c	*No. Completed* n
1	29	4
2	20	3
3	6	3
4	24	6
5	13	4
6	18	5
7	9	2
8	7	3
9	16	4
10	3	1

$$UCL = \bar{u} + \frac{3\sqrt{\bar{u}}}{\sqrt{n}}$$

$$UCL = 4.14 + \frac{3\sqrt{4.14}}{\sqrt{4}}$$

$$UCL = 7.19$$

$$LCL = \bar{u} - \frac{3\sqrt{\bar{u}}}{\sqrt{n}}$$

$$LCL = 4.14 - \frac{3\sqrt{4.14}}{\sqrt{4}}$$

$$LCL = 1.09$$

The results of calculations for the remaining days are in Table 8.8.

Answer:

Figure 8.5.

Table 8.8 Average Number of Defects (u) and Limits

DAY	*c*	*n*	*u*	*UCL*	*LCL*
1	25	4	6.25	7.19	1.09
2	20	3	6.67	7.66	0.62
3	6	3	2.00	7.66	0.62
4	24	6	4.00	6.63	1.65
5	13	4	3.25	7.19	1.09
6	18	5	3.60	6.87	1.41
7	9	2	4.50	8.46	0.00
8	7	3	2.33	7.66	0.62
9	16	4	4.00	7.19	1.09
10	3	1	3.00	10.24	0.00

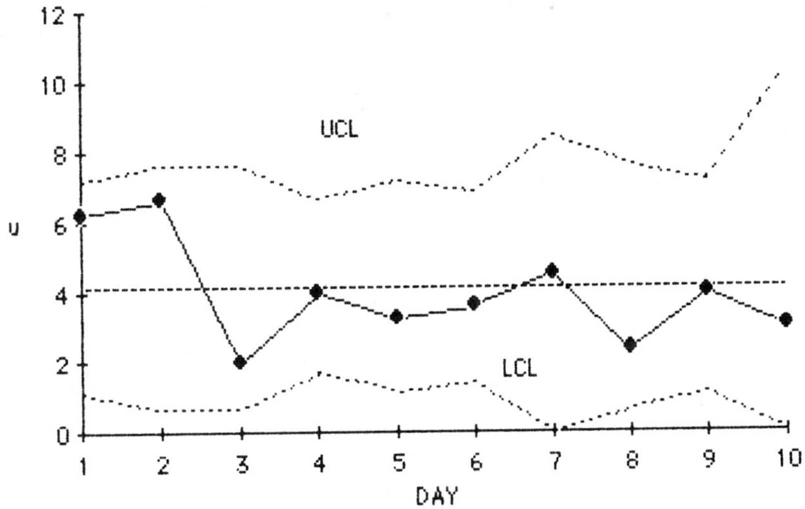

Figure 8.5 u Chart

Similarity of np and c Charts

The np and c charts are plots of integer values. The equations for the three-sigma limits are similar.

$$UCL_c = c' + 3\sqrt{c'}$$

$$UCL_{np} = n\bar{p} + 3\sqrt{n\bar{p}(1 - \bar{p})}$$

When p is small, say 0.1, and the value of c′ or np is five or more, both equations give approximately the same control-limit values. In this case,

$$UCL_c = 11.7$$

$$UCL_{np} = 11.4$$

Since the plotted values are integers, both place a value below 11 in control and a value of 12 out of control.

We will use this similarity of the binomial and Poisson distributions to simplify the calculations of confidence levels for detection of production quality shifts. The slight error introduced by this simplification is probably no greater than that introduced by using the binomial and Poisson distributions to estimate the actual population distribution.

Detecting Shifts in Average Quality

The p chart is most frequently used in conjunction with 100% inspection. The sample sizes are large and p can have many values. Since there is already 100% inspection, the sample size cannot be increased. Since p can have many values, the probability of detecting a mean shift can be increased by using two-sigma control limits rather than three-sigma limits.

The c and np Charts

The similarity of the binomial and Poisson distributions is shown in Figure 8.6. The binomial distribution is for n = 30 and p′ = 0.222, giving np = 6.66. The Poisson distribution is for c′ = 6.66. There are unlimited combinations of p′ and n that will give a specific value to np. To simplify the calculation of confidence levels for detecting population shifts, we will develop one system for c′ and use the value of np as equal to c′.

Detecting Population Shifts

Figure 8.7 shows the shift of a distribution to the right as the value of c′ or np goes from 6 to 7 to 8 and to 9.

In Example 8-3 we calculated 14.3 as the UCL_{np} for a population with an np of 6.66, n = 300, p′ = 0.0222. The probability of detecting a population shift to a p′ = 0.03 which would make np = 9 is the area under the right-most curve that corresponds to the probability of r being 14 or more.

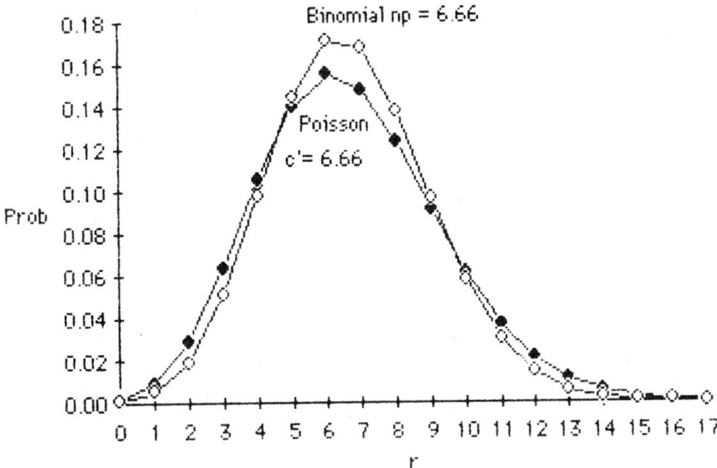

Figure 8.6 Binomial and Poisson Distributions

Figure 8.8 shows the cumulative probability for populations with a c′ or np of 6, 7, 8, and 9 having r or less defects in a single item or sample. From this chart we can see that a population with a c′ of 6 would have a 0.999 probability of 14 or less defects in a single item. For np = 6.66 it would be about 0.998.

The same chart shows that a sample from a population with an np′ of 9 would have a 0.9 probability of having 14 or fewer defects. This gives us a 0.1 probability of detecting a population shift on a single sample with an $UCL_{np} = 14+$.

With the $\overline{X}$ chart we could narrow the control chart limits by increasing the sample size because $\sigma_{\overline{X}} = \sigma'/\sqrt{n}$. For the np chart the square root of the sample size is in the numerator of the standard deviation formula, and an increased sample size has less effect. If we triple the sample size to 900 in Example 8-3, np = 19.98 and the upper control limit becomes $UCL_{np} = 33.2$. If the population p′ shifts to 0.03, np becomes 27.

Appendix A, Table C contains a list of the cumulative probabilities for the Poisson distribution. It shows, for a distribution with a c′ or np of 27, that there is a 0.89 probability of a sample having 33 or less defects or defectives. This gives us a 0.11 probability of detecting the population shift in a single sample. Further examination of Table C shows that for c′ or np = 27, the probability of a sample being plotted below the np chart centerline, 19.9, is only 0.07. It is most likely that the population shift will first be detected by the observation of several plotted points being above the centerline.

Where the quality level is high, it takes very large sample sizes to detect minor changes in the population.

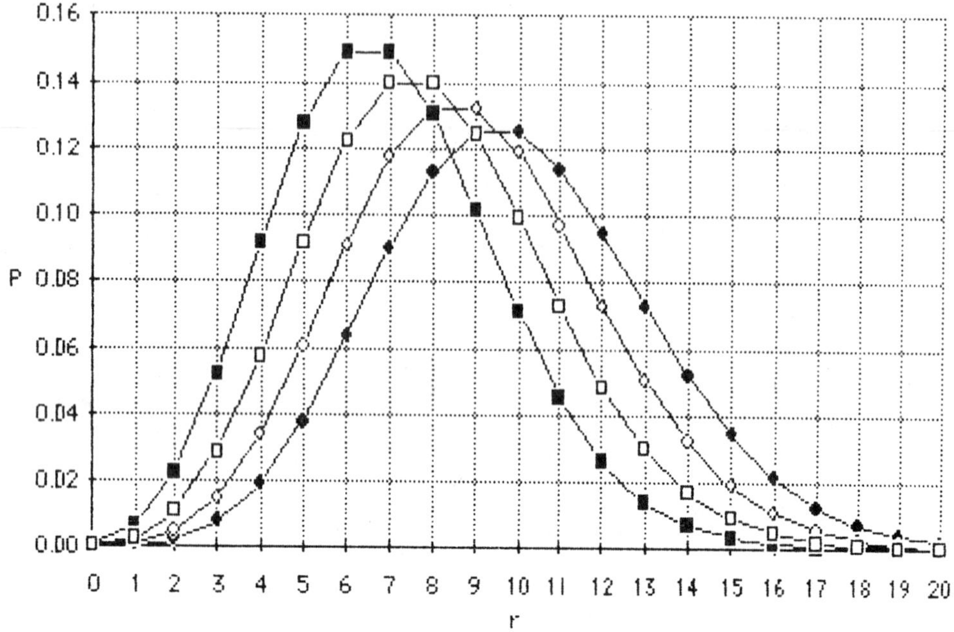

Figure 8.7 Poisson Distributions

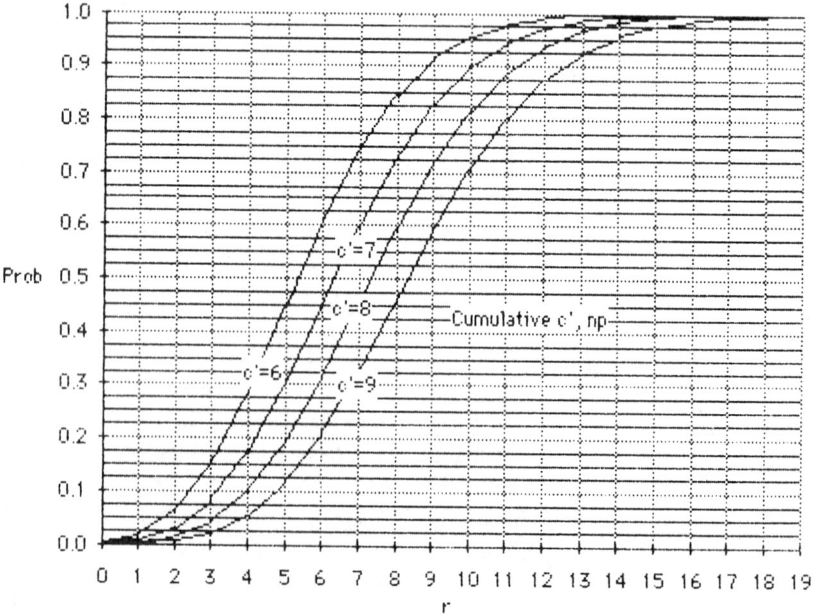

Figure 8.8 Cumulative Poisson Distributions

Example 8-6

Given:

A population has a p′ = 0.10. An np control chart is being used with a sample size of 100.

Find:

a. The probability of detecting a population quality shift to 0.13 with the current chart.
b. The probability of detecting this change with a sample size of 200.
c. What is the probability of a sample of 100 from the new population having 10 or less defective items?
d. What is the probability of a sample of 200 from the new population having 20 or less defective items?

Solution:

a.

$$UCL_{np} = n\bar{p} + 3\sqrt{n\bar{p}(1 - \bar{p})}$$

$$UCL_{np} = 10 + 3\sqrt{10(1 - 0.10)}$$

$$UCL_{np} = 19$$

For p′ = 0.13, np = 13. From Table C, a population with np = 13 has a 0.957 probability of 19 or less defective items. The probability of detection is 1 − 0.957 = 0.043.

b.

$$UCL_{np} = 20 + 3\sqrt{20(1 - 0.10)}$$

$$UCL_{np} = 32.7$$

For p′ = 0.13, np = 26. From Table C, a population with np = 26 has a 0.896 probability of 32 or less defective items. The probability of detection is 1 − 0.896 = 0.104.

c. From Table C a population with np = 13 has a 0.252 probability of 10 or less defective items.

d. From Table C, a population with np = 26 has a 0.139 probability of 20 or less defective items.

Answer:

a. P(detecting p′ of 0.13, n = 100) = 0.043
b. P(detecting p′ of 0.13, n = 200) = 0.104
c. P(10 or less defectives) = 0.252
d. P(20 or less defectives) = 0.139

Uses of the c and np Charts

With 100% inspection, these charts indicate the overall quality of the product. They will detect slight changes in quality over time and will quickly uncover significant quality changes.

Where a certain number of defects per unit are acceptable, periodic sampling will indicate overall quality with less than 100% sampling.

The charts will provide statistical data to evaluate the results of changed production procedures or worker training programs.

The Midas Inspection System is a data processor which compares the measurement results from dedicated gauging setups to user-established tolerance limits for part acceptance. The system displays acceptance or rejection of a part by colored LEDs on the front panel, indicating simultaneously the current status of 20 different inspection checks.

The part inspection results can be used as an indicator for the condition of the machining operations that produced the parts. Built-in statistical routines perform analysis of gauging results by lot. Additional statistical features may be selected for lot-to-lot comparisons. Display devices include a CRT and a printer.

Chapter Review

Formulas

Limits for p charts:

$$UCL_p = \bar{p} + 3\sqrt{\frac{\bar{p}(1-\bar{p})}{n}}$$

$$LCL_p = \bar{p} - 3\sqrt{\frac{\bar{p}(1-\bar{p})}{n}}$$

Limits for np charts:

$$UCL_{np} = n\bar{p} + 3\sqrt{n\bar{p}(1-\bar{p})}$$

$$LCL_{np} = n\bar{p} - 3\sqrt{n\bar{p}(1-\bar{p})}$$

Limits for c charts:

$$UCL_c = c' + 3\sqrt{c'}$$

$$LCL_c = c' - 3\sqrt{c'}$$

Limits for u charts:

$$UCL = \bar{u} + \frac{3\sqrt{\bar{u}}}{\sqrt{n}}$$

$$LCL = \bar{u} + \frac{3\sqrt{\bar{u}}}{\sqrt{n}}$$

Problems

8-1 The estimated quality of a product has been $p' = 0.0118$. The data from the last 10 day's production is in Table 8.9. Plot the data on a p chart. Calculate and use limits based on each day's sample size, n.

8-2 The estimated quality of a product has been $p' = 0.2$. The data from the last 10 days' production is in Table 8.10. Plot the data on a p chart. Calculate and use limits based on an average sample size of 350 items.

8-3 In the chart developed for Problem 8-2, the data for day 2 and day 3 are above or near the UCL. Calculate the upper control limits for these days and determine whether or not the points are actually out of control.

8-4 Table 8.11 contains data from 15 samples of 500 items. The average quality of the items has been $p' = 0.023$. Develop an np chart and plot the data.

8-5 Table 8.12 contains data from 15 samples. The average quality of the items has been $c' = 7$. Develop a c chart and plot the data.

8-6 Table 8.13 contains data from 15 samples of 100 items. The average quality of the items has been $p' = 0.087$. Develop an np chart and plot the data.

8-7 Table 8.14 contains data from 15 samples of 100 items. The average quality of the items has been $p' = 0.045$. Develop an np chart and plot the data. Is this process under control?

8-8 Calculate the probability that the control chart in Problem 8-4 will detect a quality shift to $p' = 0.05$ in a single sample.

8-9 Calculate the probability that the control chart in Problem 8-5 will detect a quality shift to $c' = 12$ in a single sample.

8-10 Calculate the probability that the control chart in Problem 8-6 will detect a quality shift to $p' = 0.15$ in a single sample.

Table 8.9 Data from 10 Days' Production

DAY	*n*	*No. Defective*
1	1562	9
2	1687	21
3	1495	19
4	1987	28
5	1520	19
6	1564	16
7	1497	21
8	1603	13
9	1486	24
10	1553	18

Table 8.10 Data for 10 Days' Production

DAY	*n*	*No. Defects*	*DAY*	*n*	*No. Defects*
1	356	75	6	389	68
2	324	89	7	355	87
3	389	100	8	340	76
4	319	60	9	321	59
5	368	53	10	376	68

Table 8.11 Data from 15 Samples

Sample Number	*No. of Defects*	*Sample Number*	*No. of Defects*
1	5	9	7
2	11	10	16
3	19	11	9
4	9	12	7
5	12	13	18
6	8	14	4
7	15	15	13
8	11		

Table 8.12 Data from 15 Samples

Item Number	*No. of Defects*	*Item Number*	*No. of Defects*
1	4	9	10
2	7	10	6
3	3	11	8
4	9	12	4
5	11	13	9
6	4	14	7
7	7	15	8
8	4		

Table 8.13 Data from 15 Samples

Sample Number	*No. of Defects*	*Sample Number*	*No. of Defects*
1	5	9	7
2	11	10	16
3	19	11	9
4	9	12	7
5	12	13	18
6	8	14	4
7	15	15	13
8	11		

Table 8.14 Data from 15 Samples

Sample Number	*No. of Defects*	*Sample Number*	*No. of Defects*
1	5	9	7
2	7	10	6
3	3	11	9
4	6	12	3
5	0	13	1
6	8	14	5
7	9	15	3
8	3		

9 Sampling Plans

This chapter covers acceptance sampling. It includes the relation between sample size and confidence of the supplier and receiver that inferior lots will be rejected and satisfactory lots will be accepted. It covers standard sampling plans and examines acceptance based on the suppliers quality assurance programs.

Objectives

On completion of this chapter you will be able to:

- develop a sampling plan.
- calculate the probability of accepting a satisfactory lot.
- calculate the probability of accepting an inferior lot.
- calculate the probability of rejecting a satisfactory lot.
- use a standard sampling plan for attributes.
- use a standard sampling plan for variables.

Acceptance Sampling

Few manufacturing organizations make all the components of their product. Most retail organizations make none or few of the products they sell. For products received from a producer, the consumer can assure quality by inspection of the product. We will call this *acceptance sampling*.

When to Inspect

Just as the use of control charts is an integral part of the manufacturing process, acceptance sampling must be an integral part of the manufacturing and retailing process. The time to assure the quality of items received from a producer is before it hurts to find the defective items. In the manufacturing or retailing process, when acceptance inspection occurs depends on the overall process.

Generally, the easiest time to do acceptance sampling is when the item is received. The major advantage is that the people doing the inspecting know that inspection is their job, and they can be trained for it. Other people in the organization know that the items have been inspected prior to them receiving the items.

In a quality retail store, however, most items are checked by the customer and the sales clerk prior to sale. Inspecting these items on receipt would be an unnecessary expense if the average quality level is high. In this case the sales clerk must be trained and understand the inspection responsibility.

In some assembly operations the best qualified person to make the acceptance test is the person installing a supplied component. This inspection responsibility must be understood and be part of the assembly operation.

It is normal to have a combination of places of inspection. A major computer manufacturer assembles 100% of a few key components in this acceptance facility and tests them under load in a hot room for three days. Other components receive sampling inspection on receipt and some are not inspected until the completed assembly gets an operational test before going to the shipping department.

The Sample

In taking samples for quality control charts, we collected the items for each sample at one time. The purpose was to get a picture of what the process was producing at that time. In acceptance sampling we want a picture of the whole lot being examined. The sample must be selected from each part of the lot.

A Single Sampling Plan

A single sampling plan is one in which the decision to accept or reject the lot is based on the result of a single sample.

If we are buying bullets by the case, carefully select a sample of 10 from different parts of the case, and successfully fire all 10 bullets, what do we know for certain? Only that we used to have 10 good bullets. Any information we infer from this inspection is based on probability. It is not probable that we selected the only 10 good bullets in the lot, but it is also not certain that all the bullets in the lot are satisfactory.

We will use the Poisson probability distribution to examine two sampling plans we might use to determine whether or not to accept a lot of 1000 bullets. Since a satisfactory bullet must meet certain velocity requirements, we expect only 99% of the lot to be satisfactory; that is, we will accept $p' = 0.01$ or 1% defectives. Our first plan is to test 10 bullets from each lot and accept the lot if all are good and reject it if one or more are defective.

We will calculate np for several possible values of p′ and determine from Appendix A, Table C, the probability that a sample with that p′ will contain zero defectives; that is, the probability that we will accept the lot (see Table 9.1).

Table 9.1 is not a very good acceptance plan. It shows that we would accept a lot containing as many as 10% defectives 36% of the time. In Example 9-1 we will repeat the calculations for a sample size of 100. Since we will accept $p' = 0.01$, we will reject the lot if the sample contains two or more defectives.

Example 9-1

Given:

One lot of 1000 bullets. $n = 100$. Accept on 1 or less defectives in the sample.

Find:

The probability of accepting lots with $p' = 0.01$ through $p' = 0.06$.

Solution:

$$For\ p' = 0.01$$

$$np = 100 \times 0.01$$

$$np = 1$$

$$P(acceptance) = P(r = 1\ or\ less)$$

$$From\ Table\ C\ for\ np = 1,\ P(r = 1\ or\ less)\ is\ 0.736$$

Answer:

Table 9.2 shows the values for $p' = 0.01$ thru 0.06.

This is a better plan than the sample size of 10. The larger sample gives a better view of the population. The producer of the bullets has some reason for concern with this larger sampling plan. We will accept a satisfactory lot with $p' = 0.01$ only 73.6% of the time.

The *producer's risk* is the probability that a satisfactory lot will be rejected. In this example it is 0.264, 1 − P(acceptance). The *consumer's risk* is the probability of

Table 9.1 Probability of Acceptance

n	p'	np'	$P(acceptance) = P(R=0)$
10	0.01	0.1	0.905
10	0.1	1	0.368
10	0.2	2	0.135
10	0.3	3	0.05
10	0.4	4	0.018

Table 9.2 Probability of Acceptance

n	p'	np'	$P(accept)$
100	0.01	1	0.736
100	0.02	2	0.406
100	0.03	3	0.199
100	0.04	4	0.093
100	0.05	5	0.04
100	0.06	6	0.017

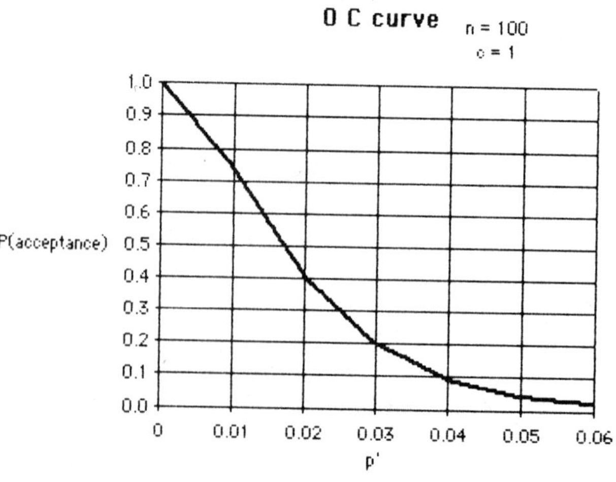

Figure 9.1 Single Sample

accepting a lot with a specific unsatisfactory quality. For example, P(accepting p′ = 0.02) is 0.402. The only sampling plan that will reduce the producer's risk *and* the consumer's risk to zero is 100% inspection. Unfortunately, when the product is consumed in the inspection, 100% inspection is not useful. In other cases, 100% inspection is neither economical nor necessary.

A graph of the probability of accepting a lot with varying values for p′ is called an *operating characteristic curve* for the sampling plan, OC curve for short. The OC curve for the sampling plan in Example 9-1 is shown in Figure 9.1.

Symbols Used in Acceptance Plans

N — Number of items in the lot being sampled.

n — Number of items in the sample.

c — Acceptance number. The maximum number of defective pieces in a sample that will allow acceptance of the lot. Also called A_c.

Pa — Probability of acceptance.

β (beta) — Consumer's risk. The probability of accepting a lot of a specific undesirably low p′

α (alpha) — Producer's risk. The probability of rejecting a lot with a satisfactory p′. $\alpha = 1 - \text{Pa}$.

A Double Sampling Plan

A double sampling plan essentially breaks a large sample into two parts. If it is highly probable from the first part of the sample that the lot meets the required quality, a decision is made to accept. If it is highly unlikely that the lot is satisfactory from the first part of the sample, it is rejected. If neither of these two conditions is met, the second part of the sample is taken and a decision made.

The additional symbols used in double sampling are:

n1 — Number of items in first sample.

c1 — Acceptance number for the first sample.

n2 — Number of items in the second sample.

c2 — Acceptance number of both samples combined.

Example 9-2

Given:

One lot of 1000 bullets is to be inspected with the following plan. A sample of 100 will be taken. The lot will be accepted if the sample contains 1 or less defectives. It will be rejected if a sample contains 3 defectives. If the sample contains 2 defectives, a second sample of 100 will be taken and the lot accepted if the second sample contains no defectives.

N = 1000

$n1 = 100$

$c1 = 1$

$n2 = 100$

$c2 = 2$

Find:

The probability of accepting lots with $p' = 0.01$ through $p' = 0.06$.

Solution:

The probability of accepting the lot on the first sample is the same as in Example 9-1. The probability of acceptance on the second sample is a compound probability. It is the probability that the first sample contained exactly 2 defectives *and* that the second sample contained zero defectives.

$$P(\textit{accept on 2nd sample}) = P(\textit{1st sample } r = 2) \times P(\textit{2nd sample } r = 0)$$

The probability of acceptance is the probability of accepting on the first sample *or* the second sample.

$$P(\textit{acceptance}) = P(\textit{accept on 1st sample}) + P(\textit{accept on 2nd sample})$$

The probability of a sample containing exactly 2 defectives can be obtained from Table C by subtracting P(1 or less) from P(2 or less). Since both samples are the same size, in this case, the value of np is the same for both samples. For $np = 1$ $P(r = 2 \text{ or less})$ is 0.92 and $P(r = 1 \text{ or less})$ is 0.736.

For $p' = 0.01$

$np = 100 \times 0.01$, $np = 1$

$P(r = 2) = P(r \leq 2) - P(r \leq 1)$

$P(r = 2) = 0.92 - 0.736$

$P(r = 2) = 0.184$

$P(r = 0)$ for $np = 1$ from Table C is 0.368

$P(\text{acc on 2nd}) = P(r = 2) \times P(r = 0)$

$P(\text{acc on 2nd}) = 0.184 \times 0.368$

$P(\text{acc on 2nd}) = 0.068$

$Pa = P(\text{a on 1st}) + P(\text{a on 2nd})$

$Pa = 0.736 + 0.068$

$Pa = 0.804$

Table 9.3 shows the values for $p' = 0.01$ thru 0.06.

Answer:

The OC curve for this sampling plan is shown in Figure 9.2. Taking the second sample reduced the producer's risk, β, from 0.26 to 0.2. The consumer has a risk, α, of 0.1 that a lot with p′ = 0.04 will be accepted.

Multiple and Sequential Sampling Plans

Multiple sampling plans are sampling plans requiring more than two samples. Sequential sampling plans have no specified number of samples. Items are inspected until a decision is reached to accept or reject the lot.

Average Outgoing Quality, AOQ

In many cases the inspection process actually improves the quality of the product accepted. In these cases rejected lots receive 100% inspection and all defective items are replaced. 100% inspection, however, cannot be used when the inspection requires a destructive test.

Table 9.3

p′	*np′*	*P(a on 1st)*	*P(r=2)*	*P(r=0)*	*P(a on 2nd)*	*Pa*
0.01	1	0.736	0.184	0.368	0.068	0.804
0.02	2	0.406	0.271	0.135	0.037	0.443
0.03	3	0.199	0.224	0.05	0.011	0.210
0.04	4	0.092	0.146	0.018	0.03	0.095
0.05	5	0.04	0.085	0.0007	0.000	0.040
0.06	6	0.017	0.045	0.002	0.000	0.017

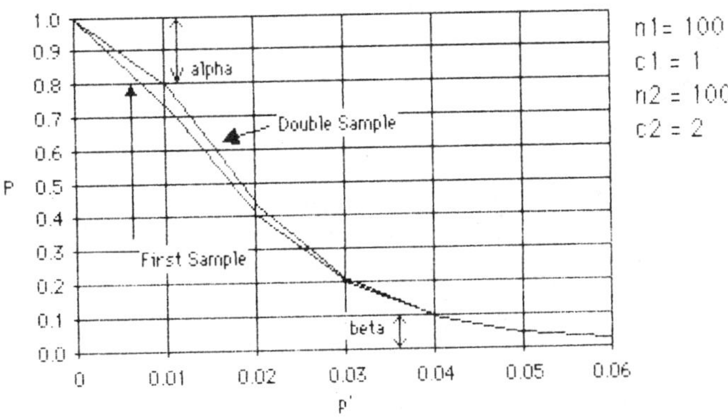

Figure 9.2 Double Sample

In cases where all defective items in rejected lots are replaced, the total number of defective items received is the number accepted in the accepted lots.

If a lot with p′ = 0.2 has a Pa of 0.60, 60% of the lots will have 20% defective items. The remaining 40% of the lots will have 0% defectives. The average quality of the lots will be:

$$0.2 \times 0.6 + 0. \times 0.4 = 0.12$$

The average outgoing quality, AOQ, for a specific value of p′ is given by the equation:

$$AOQ = p' \ Pa$$

The maximum AOQ value for a sampling plan is called the *average outgoing quality limit*, AOQL. It is the poorest average quality that would be received under a given sampling plan where all defective items in rejected lots are replaced with satisfactory items.

Example 9-3

Given:

A sampling plan has been proposed to accept ceramic capacitors. It is a single sampling plan for accepting lots of 3000 items. The overall desired quality is p′ = 0.01. The sample size is 400 with an acceptance number of 6.

Find:

The plan AOQL, the consumer's risk for p′ = 0.025, and the producer's risk for p′ = 0.01.

Solution:

$$N = 3000$$

$$n = 400$$

$$c = 6$$

$$For\ p' = 0.01$$

$$np = 400 \times 0.01 = 4$$

From Chart C for np = 4 and r = 6 or less, Pa = 0.889

$$AOQ = p' \ Pa$$

$$AOQ = 0.01 \times 0.889$$

$$AOQ = 0.0089$$

The data for p′ = 0.01 to 0.035 is in Table 9.4.

Table 9.4

p'	np'	Pa	AOQ
0.010	4	0.889	0.0089
0.015	6	0.606	0.0091
0.020	8	0.313	0.0063
0.025	10	0.130	0.0033
0.030	12	0.046	0.0014
0.035	14	0.014	0.0005

The AOQL is the maximum AOQ value, 0.0091.

The consumer's risk, β, for $p' = 0.025$ is 0.13

The producer's risk is:

$$\alpha = 1 - Pa \, for \, p' = 0.01$$
$$\alpha = 1 - 0.889$$
$$\alpha = 0.111$$

Answer:

$AOQL = 0.0091.$

$\beta, for \, p' = 0.025 \, is \, 0.13$

$\alpha = 0.111$

Standard Sampling Plans

We simplified the actual calculations for the OC curves because the Poisson distribution is close enough to the binomial for quality assurance sampling. Even with that simplification, it is a difficult process to calculate the many possible sampling plans and then select the most economical. Almost all sampling plans in use are taken from plans already developed.

In its book, *Profit Through Quality: Quality Assurance Programs for Manufacturers*, the Quality Control and Reliability Engineering Division of the American Institute of Industrial Engineers recommends the following Defense Department plans:

MIL - HDBK - 53 -1A "Guide for Attribute Lot Sampling Inspection and MIL - STD - 105"

MIL - STD - 105D	"Sampling Procedures and Tables for Inspection by Attributes"
MIL - STD - 414	"Sampling Procedures and Tables for Inspection by Variables for Percent Defective"

These publications are available from The Naval Publications and Forms Center, 5801 Tabor Ave., Philadelphia, PA 19120. MIL - STD - 105D has been republished as ANSI/ASQC Z1.4-1981 American National Standard "Sampling Procedures and Tables for Inspection by Attributes." It is published by the American Society for Quality Control, 230 West Wells Street, Milwaukee, WI 53203.

Selecting a Standard Sampling Plan

The primary characteristic for selecting a standard sampling plan is the acceptable quality level, AQL. The AQL is the maximum percent defective (or the maximun number of defects per 100 items where more than one defect per item is acceptable) that can be considered satisfactory as a process average. In Example 9-2 the AQL was 0.01 or 1 %. The standard sampling plans provide the sample size based on lot size and AQL. The plans also provide complete OC curve data and AOQL. The MIL STD plans also include double and multiple sampling plans which best fit the OC curve of a given single sampling plan, along with charts showing the average sample size for these plans. The OC curves for AQLs of 10% or less and sample sizes of 80 or less are based on the binomial distribution. The remainder are based on the Poisson distribution. Extracts of MIL-STD - 105D are contained in Appendix C.

Example 9-4

Given:

A lot of 1000 capacitors is to be inspected. AQL is to be 0.01.

Find:

a. A single sampling plan from MIL - STD - 105D.
b. The plan AOQL.
c. The consumer's risk for $p' = 0.025$.
d. The producer's risk for $p' = 0.01$.
e. Alternate double and multiple plans.
f. The average sample size for the alternate plans.

Solution:

TABLE I, (Figure 9.3), provides a sample size code letter for various lot sizes. For a lot size 501 to 1200, lot size J is specified for general inspection level II, the level for normal use.

TABLE II-A, (Figure 9.4), provides the sample size, acceptance number Ac, rejection number Re, for single sampling plans and normal inspection. For sample size J and AQL = 1%, the sample size is 80, the Ac is 2, and Re is 3.

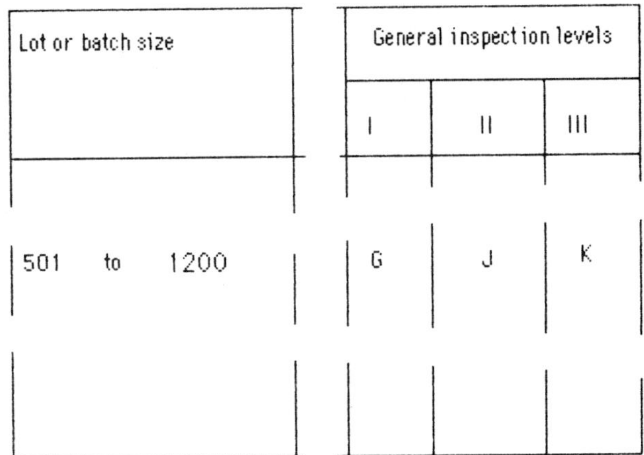

Lot or batch size	General inspection levels		
	I	II	III
501 to 1200	G	J	K

Figure 9.3 TABLE 1-Sample size code letters

Sample size code letter	Sample size	Acceptable Quality Levels			
			1.0		
			Ac	Re	
J	80		2	3	

Figure 9.4 TABLE II-A, Single sampling plans

TABLE V-A, (Figure 9.5), provides the AOQL factor of 1.7.

AOQL = factor [1 − n/N] = 1.7 [1 − 80/1000] = 1.56

TABLE X-J provides the OC curve and tabulated values. It shows Pa of 95% for p′ = 0.01 or 1% and a Pa of 70% for p′ = 0.025 or 2.5%.

$\alpha = 1 - Pa = 1 - 0.95$. $\alpha = 0.05$

Code letter	Sample size	Acceptable Quality Level		
			1.0	
J	80		1.7	

Figure 9.5 TABLE V-A, Average Outgoing Quality Limit Factors

TABLE III-A provides double sampling plans for normal inspection. For sample size J and AQL = 1%:

$n1 = 50$	$Ac1 = 0$	$Re1 = 3$
$n2 = 50$	$Ac2 = 3$	$Re2 = 4$

TABLE IV-A provides multiple sampling plans for normal inspection. For sample size J and AQL = 1%:

$n1 = 20$	*none*	$Re1 = 2$
$n2 = 20$	$Ac2 = 0$	$Re2 = 2$
$n3 = 20$	$Ac3 = 0$	$Re3 = 2$
$n4 = 20$	$Ac4 = 1$	$Re4 = 3$
$n5 = 20$	$Ac5 = 2$	$Re5 = 3$
$n6 = 20$	$Ac6 = 3$	$Re6 = 3$
$n7 = 20$	$Ac7 = 4$	$Re7 = 3$

TABLE IX provides the average sample size for double and multiple sampling plans. The vertical axis refers to fractions of the single sample size. The horizontal axis is in np. A separate chart is given for single sampling c or Ac numbers from 1 to 44. For c = 2, a double sampling plan has an average sample size that is approximately 80% of the single sample plan. The multiple plan has an average sample size of approximately 70% of the single sample plan.

Answer:

a. Single sampling plan: n = 80, c = 2
b. AOQL = 1.56%
c. Pa of 70% for p′ = 0.025
d. $\alpha = 0.05$
e. Double sampling plan

$n1 = 50$	$Ac1 = 0$	$Re1 = 3$
$n2 = 50$	$Ac2 = 3$	$Re2 = 4$

Multiple sampling plan

$n1 = 20$	*none*	$Re1 = 2$
$n2 = 20$	$Ac2 = 0$	$Re2 = 2$
$n3 = 20$	$Ac3 = 0$	$Re3 = 2$
$n4 = 20$	$Ac4 = 1$	$Re4 = 3$
$n5 = 20$	$Ac5 = 2$	$Re5 = 3$
$n6 = 20$	$Ac6 = 3$	$Re6 = 3$
$n7 = 20$	$Ac7 = 4$	$Re7 = 3$

f. Average sample size
Double sampling plan 64
Multiple sampling plan 56

This inspection plan also provides for reduced sampling when the producer maintains high quality and increased sampling when poor quality causes several lots to be rejected.

Sampling for Variables

Sampling for acceptance of items which must meet measurable specifications can be done in two ways.

- The item can be compared to the specification limits and accepted or rejected as with attributes.
- The item can be measured and the sample distribution used to estimate the proportion of the population outside the specifications.

The simple accept-reject inspection requires larger samples but the inspection is quicker. For dimensions, simple go-nogo gages can be used with less inspector training. An attribute sampling plan is used.

Using the sample distribution requires the measurement of each item and a combination of calculations and Tables found in MIL-STD-414. This publication contains several examples to demontrate the procedures it contains. The following example is taken from it.

Example 9-5

The minimum temperature of operation for a certain device is specified as 180°F. The maximum temperature is 209°F. One lot of 40 items is submitted for inspection. Inspection level IV, normal inspection, with AQL = 1% is to be used. Tables A-2 and B-3 (Appendix C) indicates that a sample size 5 is required. Suppose the measurements obtained are as follows: 197°, 188°, 184°, 205°, and 201°; and compliance with the acceptability criterion is to be determined.

Line	Information Needed	Value Obtained	Explanation
1	Sample size: n	5	Tables A-2 and B-3
2	Sum of measurements: ΣX	975	
3	Sum of squared measurements: ΣX^2	190435	
4	Correction Factor, CF: (ΣX^2)/n	190125	$(975)^2/5$
5	Corrected Sum of Squares: $\Sigma X2 - CF$	310	190435-190125
6	Variance, V ($\Sigma X^2 - CF)/(n-1)$	77.5	310/4
7	Est. Lot Standard Deviation, s: $\sqrt{V}$	8.81	$\sqrt{77.5}$
8	Sample Mean $\bar{X}$: $\Sigma X/n$	195	975/5
9	Upper Specification Limit: U	209	
10	Lower Specification Limit: L	180	
11	Quality Index: $Q_U = (U - \bar{X})/s$	1.59	(209-195)/8.81
12	Quality Index: $Q_L = (\bar{X} - L)/s$	170	(195-180)/8.81
13	Est. Lot % Def. above U: P_U	2.19%	See Table B-5
14	Est. Lot % Def. below L P_L	0.66%	See Table B-5
15	Tot. Est. % Def.: $p = P_U + P_L$	2.85%	2.19% +0.66%
16	Max. Allowable % Def.: M	3.32%	See Table B-3
17	Acceptability Criterion:	2.85% <3.32%	

Compare p with M.

The lot meets the acceptability criterion since p is less than M.

A Complete Acceptance Sampling Procedure

The following are extracts from the sampling procedure used by L. L. Bean, Inc. They are reproduced with their permission. Their overall plan requires that every item received for sale through their catalog pass acceptance sampling. The inspectors are thoroughly trained. For each type item there is a complete check list showing what is to be inspected and criterion for acceptance. Samples of items are on hand for comparison.

Most of the quality is designed into the items on the basis of field tests and laboratory analysis. The acceptance plan assures that the manufacturers, including company owned plants, comply with all the specifications.

Defects

Only major defects will be considered in application of the sampling plan.

A major defect is defined as a defect which would render the product to less than full value. The consumer would consider the product to be less than full value. The consumer would be unwilling normally to accept such a defect at full retail price. Any defect which is likely to reduce materially the useability of the product for its intended use, or is likely to result in failure is also a major defect.

A minor defect is defined as one which if noticed by the consumer would not cause any objection. The value of a product is not affected by the presence of a minor defect. The defect would not reduce materially the useability of the product for its intended use, nor is it likely to result in failure. Minor defects are not considered in application of the attached sampling program.

A critical defect is one that is likely to result in injury or unsafe conditions. Such a defect is not considered in application of this plan (normally any product with a critical defect would be given special priority and consideration).

Lot Size

The lot size is the amount received in one shipment from the vendor. This lot will consist of one style, but can include all colors and sizes received from one vendor in one shipment.

Sub-Lots

When deemed necessary, additional information may be needed concerning portions of the total lot. These will be called sub-lots and may consist of a seperate color of a style within the lot, or even a seperate size.

Plan

The plan to be used in acceptance sampling at L. L. Bean will be the Double Sampling Plan. This plan requires less sampling normally than a Single Plan, and is less complicated to administer than is the Multiple Plan.

Procedure

Using the attached plan, Table I, determine the lot size in units. Select the first sample *at random* from the shipment. Keep the proportion of sizes/colors in the sample the same as the proportion of size/color in the shipment. Other than being proportioned as noted, the sample must be a random one to prevent possible bias. Inspect the first sample lot as selected. Acceptance or rejection can be determined by comparing the number of items with major defects as defined with the A (Accept) and R (Reject) columns. In the event the number of defective units found are between the A and R figures in the table, a second sample is to be selected in the same manner as before. The cumulative results of the first and second sample will always give a positive accept/reject decision.

Normal, Reduced, and Tightened Inspection

Normal Inspection

It will be used for inspection of original submissions from any vendor. Normal inspection will also be used when a vendor changes place of manufacture, materially alters the product, or an extended period of time (six months or over) elapses between receipts. (USE TABLE I)

Normal to Tightened

When normal inspection is in effect, tightened inspection will be instituted when two consecutive lots have been rejected on normal inspection (re-submitted lots to be ignored in applying this rule). (USE TABLE II)

Tightened to Normal

When tightened inspection is in effect, normal inspection will be instituted when two consecutive lots have been accepted on tightened inspection. (USE TABLE I)

Normal to Reduced

When normal inspection is in effect, reduced inspection will be instituted when three consecutive lots have been accepted on normal inspection. (USE TABLE III)

TABLE I Double Sampling Plan - Normal Inspection

Lot Size	*Sample*	*Cumulative Sample*	*Accept (A)*	*Reject (R)*
2-8	2	2	0	1
9-15	2	2	–	1
	2	4	0	1
16-25	3	3	–	1
	3	6	0	1
26-50	5	5	0	2
	5	10	1	2
51-90	8	8	0	2
	8	16	1	2
91-150	13	13	0	3
	13	26	3	4
151-280	20	20	1	4
	20	40	4	5
281-500	32	32	2	5
	32	64	6	7
501-1200	50	50	3	7
	50	100	8	9
1201-3200	80	80	5	9
	80	160	12	13
3201-1000	125	125	7	11
	125	250	18	19

TABLE II Double Sampling Plan - Tightened Inspection

Lot Size	*Sample*	*Cumulative Sample*	*Accept (A)*	*Reject (R)*
2-8	2	2	0	1
9-15	2	2	–	1
	2	4	0	1
16-25	3	3	–	1
	3	6	0	1
26-50	5	5	0	1
	5	10	1	1
51-90	8	8	0	2
	8	16	1	2
91-150	13	13	0	2
	13	26	1	2
151-280	20	20	0	3
	20	40	3	4
281-500	32	32	1	4
	32	64	4	5
501-1200	50	50	2	5
	50	100	6	7
1201-3200	80	80	3	7
	80	160	11	12
3201-1000	125	125	6	10
	125	250	15	16

Reduced to Normal

When reduced inspection is in effect, normal inspection will be instituted when:

a. A lot is rejected
b. Inspection under reduced inspection is terminated without either acceptance or rejection having been met. In this instance, accept the lot but reinstitute normal inspection, starting with the next lot. (USE TABLE I)

TABLE III Double Sampling Plan - Reduced Inspection

Lot Size	*Sample*	*Cumulative Sample*	*Accept (A)*	*Reject(R)*
2-8	1	1	0	1
9-15	2	2	0	1
16-25	3	3	0	1
26-50	2	2	0	2
	2	4	0	2
51-90	3	3	0	2
	3	6	0	2
91-150	5	5	0	3
	5	10	0	4
151-280	8	8	0	4
	8	16	1	5
281-500	13	13	0	4
	13	26	3	6
501-1200	20	20	1	5
	20	40	4	7
1201-3200	32	32	2	7
	32	64	6	9
3201-1000	50	50	3	8
	50	100	8	12

Sub-lots

Examination of the inspection results obtained in the sampling procedure above may indicate the need for further investigation. When more than one size or color or both are included in a main lot, it is possible that only one of these size/color combinations differ. Further inspection could then reveal that:

a. A sub-lot may be acceptable even though the main lot is to be rejected, or
b. A sub-lot may be unacceptable even though the entire lot may be accepted.

In an accepted lot, if there is in any sub-lot a case where the number of defective units exceeds 10% of the number sampled in that sub-lot, then treat the sub-lot as a main lot. Further sample if necessary to make determination of disposition. See Example 1.

In a rejected lot, if there is in any sub-lot a case where the number of defective is less than 10% of the number sampled in that sub-lot, then treat the sub-lot as a main lot. Further sample if necessary to make determination of disposition. See Example 2.

EXAMPLE 1

Lot consisting of four colors is received from one vendor. Total shipment is 1000 units. Sample is 50 (see Table I).

Color	A	100 Units	Sample 5	Reject 1
	B	300 Units	Sample 15	Reject 0
	C	200 Units	Sample 10	Reject 0
	D	400 Units	Sample 20	Reject 1
TOTAL		1000 Units	Sample 50	Reject 2

Total lot is accepted since you accept with three or less defectives from 50 samples. Yet color A with one defective unit out of five (20%) may need further consideration. Since this is a sub-lot of 100 units, go to that plan and further sample. Since it requires a sample of 13, 8 additional samples should be selected.

EXAMPLE 2

Assume the same shipment with a different reject result.

Color	A	100 Units	Sample 5	Reject 1
	B	300 Units	Sample 15	Reject 3
	C	200 Units	Sample 10	Reject 0
	D	400 Units	Sample 20	Reject 4
TOTAL		1000 Units	Sample 50	Reject 8

This lot is rejected with 8 defectives out of 50 sampled. In this instance, however, we may be able to accept sub-lot C since defectives found were less than 10% (0%). Again check the table for sample size for a sub-lot of 200 units. Since this is 20, sample an additional 10 and make a decision. It is possible, of course, that a second sample will be needed since this is a Double Sampling Plan.

It is noted that this treatment of sub-lots not only prevents acceptance of a particular color/size which is not acceptable, but also prevents the rejection of a sub-lot which can be accepted for stock if it meets the requirements.

Re-Submitted Lots

Lots or sub-lots returned to the supplier will be expected to have 100% inspection performed by that supplier. All defective units will be repaired, cleaned, or otherwise corrected, or will not be included in a resubmitted lot of first quality merchandise. Re-submitted lots will be inspected in the same manner as original lots.

Alternatives to Acceptance Inspection

Inspection costs money whether it is done in the course of manufacture or in acceptance. If the manufacturer has a good inspection plan, a second inspection by the consumer may not be necessary. Many organizations, particularly U. S. Government organizations, are including an inspection requirement in their procurement contracts. This requires the manufacturer to use a mutually agree-able inspection plan in the course of manufacture and little or no acceptance inspection is done by the consumer.

This reduces the total inspection cost. It has been less than 100% successful. In 1984 a major electronic component supplier was found to have omitted the 100% inspection called for in its contracts.

There are, however, an increasing number of manufacturing operations in which automated inspection is being used. The decision on accepting this in lieu of acceptance sampling is an economic one. If the cost of using or selling a defective product is high, acceptance sampling is necessary.

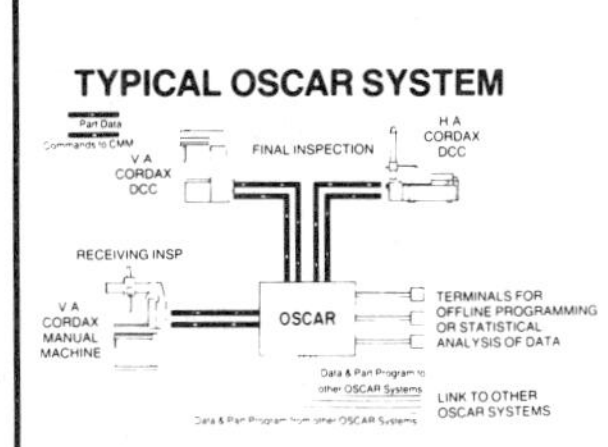

Sheffield has added to this system its OSCAR (Operating System for CMMs with Advanced Reporting) management system which allows the part designer to also pass to a CMM the quality assurance measurement instructions. OSCAR is executed through a real time, multi-tasking host computer to operate an unlimited number of CMMs simultaneously, log the collected data, and let the users at terminal workstations perform off-line programming or statistical analysis.

Chapter Review

Keywords

Producer's risk, α The probability that a satisfactory lot will be rejected.

Consumer's risk, β The probability that an unsatisfactory lot of specific poor quality will be accepted.

Operating characteristic curve, OC curve A graph of a sampling plan's probability of acceptance vs p′ of submitted lots.

Average outgoing quality The average quality received from an inspection plan after all defective items in rejected lots have been replaced.

Average outgoing quality limit, AOQL The lowest average quality received from an inspection plan after all defective items in rejected lots have been replaced.

Acceptable quality level The maximum percent defective or the maximum number of defects per 100 items that can be considered satisfactory as a process average.

Symbols

N — Number of items in the lot being sampled.

n — Number of items in the sample.

c — Acceptance number. The maximun number of defective pieces in a sample that will allow acceptance of the lot. Also called A_c.

Re — Rejection number. The minimum number of defective items in a sample that will cause rejection of the lot.

Pa — Probability of acceptance.

β (beta) — Consumers risk. The probability of accepting a lot of a specific undesirably low p′.

α (alpha) — Producer's risk. The probability of rejecting a lot with a satisfactory p′.

nl — Number of items in first sample.

cl — Acceptance number for the first sample.

n2 — Number of items in the second sample.

c2—Acceptance number of both samples combined

Formula

$\alpha = 1 - Pa$

$AOQ = p'Pa$

Problems

9-1 Calculate the probability of acceptance for lots containing from 2% to 12% defectives with a single sampling plan using n = 50 and c = 1.

9-2 Plot the OC curve for an acceptance plan using single samples of 200 and c = 3. Calculate the AOQ for each value of p′ and identify the AOQL. What is the producer's risk of submitting lots with p′ = 0.01 and the consumers risk of accepting lots with p′ = 0.03.

9-3 Plot the OC curve for an acceptance plan using two samples of 200 and c = 5. Calculate the AOQ for each value of p′ and identify the AOQL.

9-4 Plot the OC curve for an acceptance plan using single samples of 100 and c1 = 3, c2 = 4. Calculate the AOQ for each value of p′ and identify the AOQL.

9-5 Plot the OC curve for an acceptance plan using single samples of 100 and c1 = 0, c2 = 1. Calculate the AOQ for each value of p′.

9-6 Plot the OC curve for an acceptance plan using single samples of 2000 and c = 10. Calculate the AOQ for each value of p′ and identify the AOQL.

9-7 Using Tables I, H, II-A, V-A, X-H, and VI-A from MIL-STD-105D, determine the following for N = 500 and AQL = 1.5%:

a. Single sample size.
b. The acceptance number.
c. The plan AOQL.
d. The alpha for lots with 1.0% defectives.
e. The percent defectives in a lot with a consumer's risk of 10%.

9-8 Using Tables I, K, II-A, V-A, X-K, and VI-A from MIL-STD-105D, determine the following for N = 2000 and AQL = 1.0%:

a. Single sample size.
b. The acceptance number.
c. The plan AOQL.
d. The alpha for lots with 1.0% defectives.
e. The percent defectives in a lot with a consumer's risk of 10%.

9-9 Using Tables from MIL-STD-105D, determine the following for N = 5000 and AQL = 1.50%:

a. Single sample size.
b. The acceptance number.
c. The plan AOQL.
d. The alpha for lots with 1.0% defectives.
e. The percent defectives in a lot with a consumer's risk of 10%.

9-10 Using Tables A-2, B-3,and B-5 from MIL-STD-414, determine the acceptability of the following sample from a lot submitted to meet the specification 1.500 ± 0.005 inch.

Sample	*Measurement*	*Sample*	*Measurement*
1	1.502	9	1.503
2	1.497	10	1.500
3	1.503	11	1.496
4	1.508	12	1.499
5	1.501	13	1.501
6	1.498	14	1.502
7	1.498	15	1.498
8	1.503		

9-11 A second lot has been received in compliance with the specifications given for Problem 9-10. The dimensions of your sample are given below. Is this lot acceptable?

Sample	*Measurement*	*Sample*	*Measurement*
1	1.501	9	1.502
2	1.497	10	1.500
3	1.503	11	1.499
4	1.501	12	1.499
5	1.501	13	1.501
6	1.498	14	1.502
7	1.498	15	1.499

10 Integrating Quality Assurance

This chapter covers the logical collection of rejection data and using it to find and correct the root cause of rejection. It includes the use of information already available in process and assembly charts.

Objectives

On completion of this chapter you will be able to:

- collect necessary rejection data.
- display rejection data.
- use an assembly chart to find the cause of rejection.
- use a process chart to find the cause of rejection.

Collecting the Data

Statistical quality assurance requires the collection of data in some meaningful form. With measurement control charts, we were able to keep track of the measurements produced and take timely action in correcting the process when the process got out of control. The same can be true of control charts for attributes. So far, we have just counted defects or defective items. It is not enough to know only the fraction of defective parts; to take corrective action we must know first what is wrong, and then what is causing it to be wrong. The most common tool for recording what is wrong is the check sheet.

Check Sheets

A simple check sheet is a list of things to examine. Inspectors need these to know what to look for when inspecting an item. If we add a place to record the reason an item is rejected, we can collect the data required to find both the cause and the most common defects. As an example, a portion of a check sheet for a retractable ball point pen is shown in Figure 10.1.

Product: Pen model 23 Date:

No. Inspected: 300 No. in lot 1000 Lot No. 501

Type	Check	No.
Color	|||	3
Retraction	𝍸		6
Won't write - ink			1
ball point	|||	3
Plastic flashing (cap)	𝍸 |||	8
Plastic flashing (bottom)	𝍸 𝍸		11
Plastic incomplete(cap)			1
Plastic incomplete (bottom)	||||	4
Clip	||	2

Figure 10.1 Check Sheet

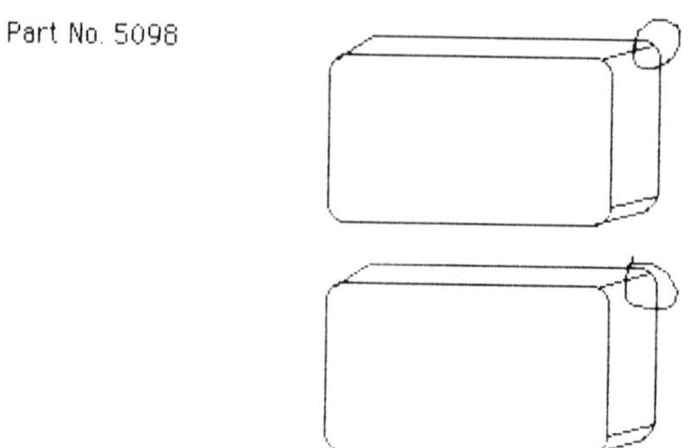

Figure 10.2 Location Check Sheet

The check sheet indicates the frequency of certain problems. It may be changed from time to time to identify specific problems or to evaluate the effect of process changes or supplier changes.

A second common form of a check sheet is a location check sheet. It contains a sketch or schematic of an item. A lubrication chart for a car is one example. It improves the quality of a lubrication operation by showing the operator the location of the many lubrication pionts.

In inspection, these are usually used to allow the inspector to indicate the location of defects. Figure 10.2 shows an example of a portion of a location check sheet.

A check sheet similar to this was used by a plastics molding company. After it was instituted, they found that almost all their rejects came from cracks in the same corner. With that information, corective action was a simple matter.

Communicating the Data

The check sheets allow for the collection of the defect data. The next step in correcting the process is to communicate with the people who know and control the process. They are interested in two things: what's the problem and how bad is it?

All problems cannot be solved at one time unless a major production change takes place. It is more common to solve problems in the order of their significance and the resources available. A common vehicle for communicating defect data is the Pareto diagram. Figure 10.3 is a Pareto diagram of the pen defects.

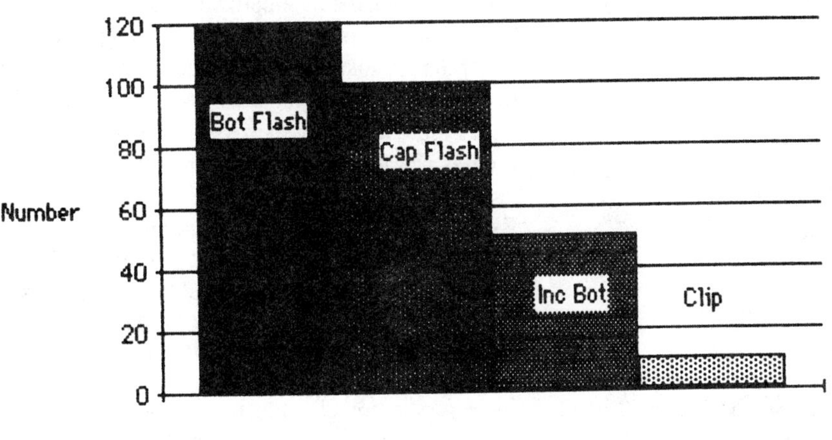

Figure 10.3 Pareto Diagram of Pen Defects

A Pareto diagram is a column chart of the significant defect data. The defect with the greatest frequency is on the left, and the chart steps down to the least frequent on the right. At a glance people can see that the most frequent defect is plastic flashing on the pen bottom and that flashing on the pen top is almost as bad. We now have shown what's wrong and how bad it is.

A second part of the "how bad is it" question has to do with the duration of the problem. Is it a constant problem or just a one-time event? Data on this can be displayed in a cumulative column chart showing the data over a period of time. Figure 10.4 shows the pen defect data over a period of two weeks. Again, the relative severity of the bottom and cap flashing problem stands out on the graph.

The Root Cause

The *root cause* of a problem is the actual cause. It may or may not be the obvious cause. Too much flashing on the pens may be caused by insufficient removal or too much coming out of the mold. Too much coming out of the mold has several possible causes.

To find the root cause, we must examine the manufacturing process and then the suspected operations.

Available Documentation

In the setting up of a manufacturing process, it is normal to develop documents which specify the manufacturing process in great detail. Formats vary, but two types are in general use: the assembly chart and the operation process chart.

The Assembly and Process Charts

The assembly and process charts are schematic diagrams showing the flow of materials from the receipt to the final product. Each operation that modified a part or material is shown. In some organizations the whole process from receipt of raw materials to packaging the product is shown on one big chart. In others, there is an overview chart showing major operations and then detailed charts of each operation. Figure 10.5 shows an overview assembly chart for a pen. Figure 10.6 shows a process chart for making the pen bottom.

Examination of the assembly and process charts shows only one inspection, the final inspection. With this quick reference it is possible to examine the possibility of inspection of the pen tops and bottoms somewhere prior to assembly, prior to the points labeled 1 in Figure 10.5. This will not sovle the problem of defective parts but at least will eliminate them prior to assembly into defective pens.

The Operation Chart

In the production planning process it is necessary to show every step in the process of making a part. The operation chart is used to determine the time, material and equipment required. This data is used to calculate the standard cost and time. It

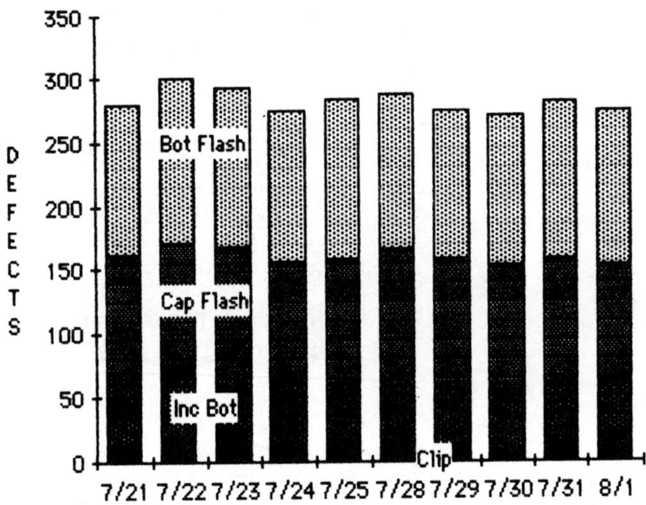

Figure 10.4 Column Graph of Defect Data Over Two Weeks

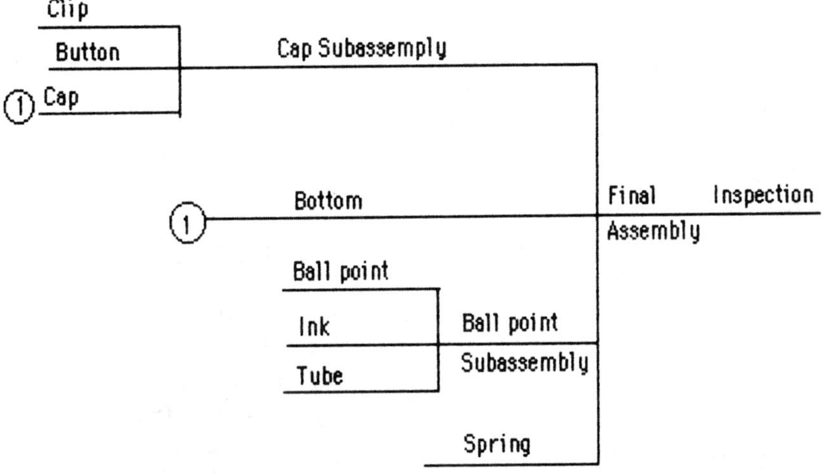

Figure 10.5 Pen Assembly Chart

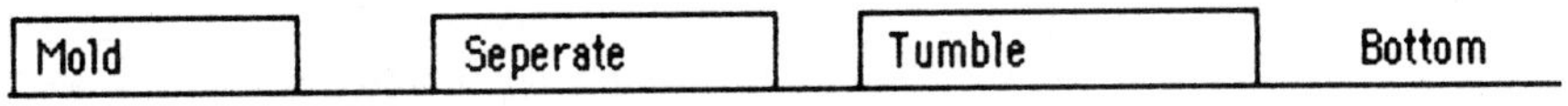

Mold | Seperate | Tumble | Bottom

Figure 10.6 Process chart, Pen bottom

Operation No. 20 Operation: Drill 19/32 in hole

Step	Details	Time
1	Place 1 part in fixture	.04
2	Clamp part	.07
3	Lower drill head	.02
4	Drill hole, 2"	.25
5	Raise head	.03
6	Release & remove part	.04
7	Blow out fixture (hose)	.05

Material: 1040 steel Machine: 24" Cinn Drill Press

Figure 10.7 Operation Chart

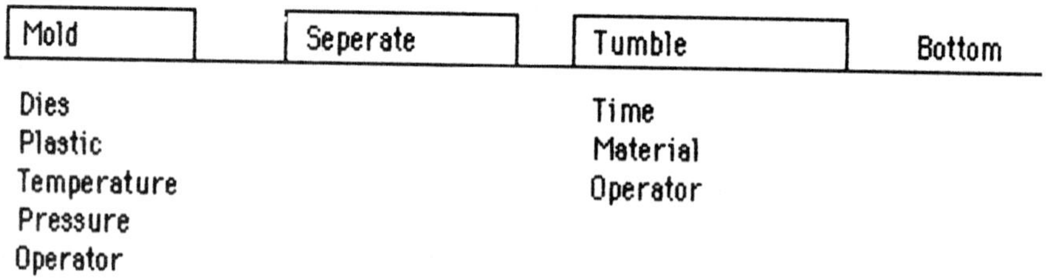

Mold	Seperate	Tumble	Bottom
Dies Plastic Temperature Pressure Operator		Time Material Operator	

Figure 10.8 Cause and Effect Diagram

shows every step performed by the worker, the machine, and the tooling. Figure 10.7 is an example of an operation chart for drilling a hole. From the information available in an operation chart and an understanding of the process involved, it is possible to develop a cause and effect diagram.

The Cause and Effect Diagram

A cause and effect diagram is a schematic diagram of a process annotated to show the possible causes of a given effect. Figure 10.8 shows Figure 10.6 modified to list the possible causes of excess flashing.

This type of chart has two advantages. The first is to assist in the mental descipline required to think of the possible causes of a specific problem. The second advantage is the communication among the people with the knowledge required to make the diagram.

Appendix A

Table A
Table B
Table C

Table A Area under the normal curve from $-\infty$ to Z

	(Numbers at column heads are second decimal value of Z)									
Z/	*0.09*	*0.08*	*0.07*	*0.06*	*0.05*	*0.04*	*0.03*	*0.02*	*0.01*	*0.00*
-3.5	0.0002	0.0002	0.0002	0.0002	0.0002	0.0002	0.0002	0.0002	0.0002	0.0002
-3.4	0.0002	0.0003	0.0003	0.0003	0.0003	0.0003	0.0003	0.0003	0.0003	0.0003
-3.3	0.0004	0.0004	0.0004	0.0004	0.0004	0.0004	0.0004	0.0005	0.0005	0.0005
-3.2	0.0005	0.0005	0.0005	0.0006	0.0006	0.0006	0.0006	0.0006	0.0007	0.0007
-3.1	0.0007	0.0007	0.0008	0.0008	0.0008	0.0009	0.0009	0.0009	0.0009	0.0010
-3.0	0.0010	0.0010	0.0011	0.0011	0.0011	0.0012	0.0012	0.0013	0.0013	0.0014
-2.9	0.0014	0.0014	0.0015	0.0015	0.0016	0.0016	0.0017	0.0017	0.0018	0.0019
-2.8	0.0019	0.0020	0.0021	0.0021	0.0022	0.0023	0.0023	0.0024	0.0025	0.0026
-2.7	0.0026	0.0027	0.0028	0.0029	0.0030	0.0031	0.0032	0.0033	0.0034	0.0035
-2.6	0.0036	0.0037	0.0038	0.0039	0.0040	0.0041	0.0043	0.0044	0.0045	0.0047
-2.5	0.0048	0.0049	0.0051	0.0052	0.0054	0.0055	0.0057	0.0059	0.0060	0.0062
-2.4	0.0064	0.0066	0.0068	0.0069	0.0071	0.0073	0.0075	0.0078	0.0080	0.0082
-2.3	0.0084	0.0087	0.0089	0.0091	0.0094	0.0096	0.0099	0.0102	0.0104	0.0107
-2.2	0.0110	0.0113	0.0116	0.0119	0.0122	0.0125	0.0129	0.0132	0.0136	0.0139
-2.1	0.0143	0.0146	0.0150	0.0154	0.0158	0.0162	0.0166	0.0170	0.0174	0.0179
-2.0	0.0183	0.0188	0.0192	0.0197	0.0202	0.0207	0.0212	0.0217	0.0222	0.0228
-1.9	0.0233	0.0239	0.0244	0.0250	0.0256	0.0262	0.0268	0.0274	0.0281	0.0287
-1.8	0.0294	0.0301	0.0307	0.0314	0.0322	0.0329	0.0336	0.0344	0.0351	0.0359
-1.7	0.0367	0.0375	0.0384	0.0392	0.0401	0.0409	0.0418	0.0427	0.0436	0.0446
-1.6	0.0455	0.0465	0.0475	0.0485	0.0495	0.0505	0.0516	0.0526	0.0537	0.0548
-1.5	0.0559	0.0571	0.0582	0.0594	0.0606	0.0618	0.0630	0.0643	0.0655	0.0668
-1.4	0.0681	0.0694	0.0708	0.0721	0.0735	0.0749	0.0764	0.0778	0.0793	0.0808
-1.3	0.0823	0.0838	0.0853	0.0869	0.0885	0.0910	0.0918	0.0934	0.0952	0.0968
-1.2	0.0985	0.1003	0.1020	0.1038	0.1057	0.1075	0.1093	0.1112	0.1113	0.1151
-1.1	0.1170	0.1190	0.1210	0.1230	0.1251	0.1271	0.1292	0.1314	0.1335	0.1357
-1.0	0.1379	0.1401	0.1423	0.1446	0.1469	0.1492	0.1515	0.1539	0.1562	0.1587
-0.9	0.1611	0.1635	0.1660	0.1685	0.1711	0.1736	0.1762	0.1788	0.1814	0.1841
-0.8	0.1867	0.1894	0.1922	0.1949	0.1977	0.2005	0.2033	0.2061	0.2090	0.2119
-0.7	0.2148	0.2177	0.2207	0.2236	0.2266	0.2297	0.2327	0.2358	0.2389	0.2420
-0.6	0.2451	0.2483	0.2514	0.2546	0.2578	0.2611	0.2643	0.2676	0.2709	0.2743
-0.5	0.2776	0.2810	0.2843	0.2877	0.2912	0.2946	0.2981	0.3015	0.3050	0.3085
-0.4	0.3121	0.3156	0.3192	0.3228	0.3264	0.3300	0.3336	0.3372	0.3409	0.3446
-0.3	0.3483	0.3520	0.3557	0.3594	0.3632	0.3669	0.3707	0.3745	0.3783	0.3821
-0.2	0.3859	0.3897	0.3936	0.3974	0.4013	0.4052	0.4090	0.4129	0.4168	0.4207
-0.1	0.4247	0.4286	0.4325	0.4364	0.4404	0.4443	0.4483	0.4522	0.4562	0.4602
-0.0	0.4641	0.4681	0.4721	0.4761	0.4801	0.4840	0.4880	0.4920	0.4960	0.5000

Table A (continued) Area under the normal curve from $-\infty$ to Z

(Numbers at column heads are second decimal value of Z)

Z/	*0.09*	*0.08*	*0.07*	*0.06*	*0.05*	*0.04*	*0.03*	*0.02*	*0.01*	*0.00*
3.5	0.9998	0.9998	0.9998	0.9998	0.9998	0.9998	0.9998	0.9998	0.9998	0.9998
3.4	0.9998	0.9997	0.9997	0.9997	0.9997	0.9997	0.9997	0.9997	0.9997	0.9997
3.3	0.9996	0.9996	0.9996	0.9996	0.9996	0.9996	0.9996	0.9995	0.9995	0.9995
3.2	0.9995	0.9995	0.9995	0.9994	0.9994	0.9994	0.9994	0.9994	0.9993	0.9993
3.1	0.9993	0.9993	0.9992	0.9992	0.9992	0.9991	0.9991	0.9991	0.9991	0.9990
3.0	0.9990	0.9990	0.9989	0.9989	0.9989	0.9988	0.9988	0.9987	0.9987	0.9986
2.9	0.9986	0.9986	0.9985	0.9985	0.9984	0.9984	0.9983	0.9983	0.9982	0.9981
2.8	0.9981	0.9980	0.9979	0.9979	0.9978	0.9977	0.9977	0.9976	0.9975	0.9974
2.7	0.9974	0.9973	0.9972	0.9971	0.9970	0.9969	0.9968	0.9967	0.9966	0.9965
2.6	0.9964	0.9963	0.9962	0.9961	0.9960	0.9959	0.9957	0.9956	0.9955	0.9953
2.5	0.9952	0.9951	0.9949	0.9948	0.9946	0.9945	0.9943	0.9941	0.9940	0.9938
2.4	0.9936	0.9934	0.9932	0.9931	0.9929	0.9927	0.9925	0.9922	0.9920	0.9918
2.3	0.9916	0.9913	0.9911	0.9909	0.9906	0.9904	0.9901	0.9898	0.9896	0.9893
2.2	0.9890	0.9887	0.9884	0.9881	0.9878	0.9875	0.9871	0.9868	0.9864	0.9861
2.1	0.9857	0.9854	0.9850	0.9846	0.9842	0.9838	0.9834	0.9830	0.9826	0.9821
2.0	0.9817	0.9812	0.9808	0.9803	0.9798	0.9793	0.9788	0.9783	0.9778	0.9772
1.9	0.9767	0.9761	0.9756	0.9750	0.9744	0.9738	0.9732	0.9726	0.9719	0.9713
1.8	0.9706	0.9699	0.9693	0.9686	0.9678	0.9671	0.9664	0.9656	0.9649	0.9641
1.7	0.9633	0.9625	0.9616	0.9608	0.9599	0.9591	0.9582	0.9573	0.9564	0.9554
1.6	0.9545	0.9535	0.9525	0.9515	0.9505	0.9495	0.9484	0.9474	0.9463	0.9452
1.5	0.9441	0.9429	0.9418	0.9406	0.9394	0.9382	0.9370	0.9357	0.9345	0.9332
1.4	0.9319	0.9306	0.9292	0.9279	0.9265	0.9251	0.9236	0.9222	0.9207	0.9192
1.3	0.9177	0.9162	0.9147	0.9131	0.9115	0.9090	0.9082	0.9066	0.9048	0.9032
1.2	0.9015	0.8997	0.8980	0.8962	0.8943	0.8925	0.8907	0.8888	0.8887	0.8849
1.1	0.8830	0.8810	0.8790	0.8770	0.8749	0.8729	0.8708	0.8686	0.8665	0.8643
1.0	0.8621	0.8599	0.8577	0.8554	0.8531	0.8508	0.8485	0.8461	0.8438	0.8413
0.9	0.8389	0.8365	0.8340	0.8315	0.8289	0.8264	0.8238	0.8212	0.8186	0.8159
0.8	0.8133	0.8106	0.8078	0.8051	0.8023	0.7995	0.7967	0.7939	0.7910	0.7881
0.7	0.7852	0.7823	0.7793	0.7764	0.7734	0.7703	0.7673	0.7642	0.7611	0.7580
0.6	0.7549	0.7517	0.7486	0.7454	0.7422	0.7389	0.7357	0.7324	0.7291	0.7257
0.5	0.7224	0.7190	0.7157	0.7123	0.7088	0.7054	0.7019	0.6985	0.6950	0.6915
0.4	0.6879	0.6844	0.6808	0.6772	0.6736	0.6700	0.6664	0.6628	0.6591	0.6554
0.3	0.6517	0.6480	0.6443	0.6406	0.6368	0.6331	0.6293	0.6255	0.6217	0.6179
0.2	0.6141	0.6103	0.6064	0.6026	0.5987	0.5948	0.5910	0.5871	0.5832	0.5793
0.1	0.5753	0.5714	0.5675	0.5636	0.5596	0.5557	0.5517	0.5478	0.5438	0.5398
0.0	0.5359	0.5319	0.5279	0.5239	0.5199	0.5160	0.5120	0.5080	0.5040	0.5000

Table B Control Chart Factors

n	c_2	d_2	A	$A1$	$A2$	$B1$	$B2$	$B3$	$B4$	$D1$	$D2$	$D3$	$D4$
2	.5642	1.128	2.121	3.760	1.880	0	1.843	0	3.267	0	3.686	0	3.267
3	.7236	1.693	1.732	2.394	1.023	0	1.858	0	2.568	0	4.353	0	2.575
4	.7979	2.059	1.5	1.88	.729	0	1.808	0	2.266	0	4.698	0	2.282
5	.8407	2.326	1.342	1.596	.577	0	1.756	0	2.089	0	4.918	0	2.115
6	.8686	2.534	1.225	1.41	.483	.026	1.711	.03	1.97	0	5.078	0	2.004
7	.8882	2.704	1.134	1.277	.419	.105	1.672	.118	1.882	.205	5.203	.076	1.924
8	.9027	2.847	1.061	1.175	.373	.167	1.638	.185	1.815	.387	5.307	.136	1.864
9	.9139	2.970	10	1.09	.337	.219	1.609	.239	1.761	.546	5.394	.184	1.816
10	.9227	3.078	.949	1.028	.308	.262	1.584	.284	1.716	.687	5.469	.223	1.777

Table C contains the Poisson probability of r or less occurrances in a population containing an average of c′. When used as an approximation of the binomial distribution, use np′ in place of c′. Where occurrances per sample are used, use nu′ in place of c′.

Table C

r	$c' = 0.1$	$c' = 1$	$c' = 2$	$c' = 3$	$c' = 4$	$c' = 5$	$c' = 6$
0	0.905	0.368	0.135	0.05	0.018	0.007	0.002
1	0.995	0.736	0.406	0.199	0.092	0.04	0.017
2	1	0.92	0.677	0.423	0.238	0.125	0.062
3	1	0.981	0.857	0.647	0.433	0.265	0.151
4	1	0.996	0.947	0.815	0.629	0.44	0.285
5	1	0.999	0.983	0.916	0.785	0.616	0.446
6	1	1	0.995	0.966	0.889	0.762	0.606
7	1	1	0.999	0.988	0.949	0.867	0.744
8	1	1	1	0.996	0.979	0.932	0.847
9	1	1	1	0.999	0.992	0.968	0,916

Table C (continued)

r	$c' = 0.1$	$c' = 1$	$c' = 2$	$c' = 3$	$c' = 4$	$c' = 5$	$c' = 6$
10	1	1	1	1	0.997	0.986	0.957
11	1	1	1	1	0.999	0.995	0.98
12	1	1	1	1	1	0.998	0.991
13	1	1	1	1	1	0.999	0.996
14	1	1	1	1	1	1	0.999
15	1	1	1	1	1	1	0.999
16	1	1	1	1	1	1	1

r	$c' = 7$	$c' = 8$	$c' = 9$	$c' = 10$	$c' = 11$	$c' = 12$
0	0.001	0	0	0	0	0
1	0.007	0.003	0.001	0	0	0
2	0.03	0.014	0.006	0.003	0.001	0.001
3	0.082	0.042	0.021	0.01	0.005	0.002
4	0.173	0.1	0.055	0.029	0.015	0.008
5	0.301	0.191	0.116	0.067	0.038	0.02
6	0.45	0.313	0.207	0.13	0.079	0.046
7	0.599	0.453	0.324	0.22	0.143	0.09
8	0.729	0.593	0.456	0.333	0.232	0.155
9	0.83	0.717	0.587	0.458	0.341	0.242
10	0.901	0.816	0.706	0.583	0.46	0.347
11	0.947	0.888	0.803	0.697	0.579	0.462
12	0.973	0.936	0.876	0.792	0.689	0.576
13	0.987	0.966	0.926	0.864	0.781	0.682
14	0.994	0.983	0.959	0.917	0.854	0.772
15	0.998	0.992	0.978	0.951	0.907	0.844

Table C (continued)

r	c' = 7	c' = 8	c' = 9	c' = 10	c' = 11	c' = 12
16	0.999	0.996	0.989	0.973	0.944	0.899
17	1	0.998	0.995	0.986	0.968	0.937
18	1	0.999	0.998	0.993	0.982	0.963
19	1	1	0.999	0.997	0.991	0.979
20	1	1	1	0.998	0.995	0.988
21	1	1	1	0.999	0.998	0.994
22	1	1	1	1	0.999	0.997
23	1	1	1	1	1	0.999
24	1	1	1	1	1	0.999
25	1	1	1	1	1	1
r	c' = 13	c' = 14	c' = 15	c' = 16	c' = 17	c' = 18
0	0	0	0	0	0	0
1	0	0	0	0	0	0
2	0	0	0	0	0	0
3	0.001	0	0	0	0	0
4	0.004	0.002	0.001	0	0	0
5	0.011	0.006	0.003	0.001	0.001	0
6	0.026	0.014	0.008	0.004	0.002	0.001
7	0.054	0.032	0.018	0.01	0.005	0.003
8	0.1	0.062	0.037	0.022	0.013	0.007
9	0.166	0.109	0.07	0.043	0.026	0.015

Table C (continued)

r	$c' = 13$	$c' = 14$	$c' = 15$	$c' = 16$	$c' = 17$	$c' = 18$
10	0.252	0.176	0.118	0.077	0.049	0.03
11	0.353	0.26	0.185	0.127	0.085	0.055
12	0.463	0.358	0.268	0.193	0.135	0.092
13	0.573	0.464	0.363	0.275	0.201	0.143
14	0.675	0.57	0.466	0.368	0.281	0.208
15	0.764	0.669	0.568	0.467	0.371	0.287
16	0.836	0.756	0.664	0.566	0.468	0.375
17	0.89	0.827	0.749	0.659	0.564	0.469
18	0.93	0.883	0.819	0.742	0.655	0.562
19	0.957	0.924	0.875	0.812	0.736	0.651
20	0.975	0.952	0.917	0.868	0.805	0.731
21	0.986	0.971	0.947	0.911	0.861	0.799
22	0.992	0.983	0.967	0.942	0.905	0.855
23	0.996	0.991	0.981	0.963	0.937	0.899
24	0.998	0.995	0.989	0.978	0.959	0.932
25	0.999	0.997	0.994	0.987	0.975	0.955
26	1	0.999	0.997	0.993	0.985	0.972
27	1	0.999	0.998	0.996	0.991	0.983
28	1	1	0.999	0.998	0.995	0.99
29	1	1	1	0.999	0.997	0.994
30	1	1	1	0.999	0.999	0.997
31	1	1	1	1	0.999	0.998
32	1	1	1	1	1	0.999
33	1	1	1	1	1	1

Table C (continued)

r	$c' = 19$	$c' = 20$	$c' = 21$	$c' = 22$	$c' = 23$	$c' = 24$
5	0	0	0	0	0	0
6	0.001	0	0	0	0	0
7	0.002	0.001	0	0	0	0
8	0.004	0.002	0.001	0.001	0	0
9	0.009	0.005	0.003	0.002	0.001	0
10	0.018	0.011	0.006	0.004	0.002	0.001
11	0.035	0.021	0.013	0.008	0.004	0.003
12	0.061	0.039	0.025	0.015	0.009	0.005
13	0.098	0.066	0.043	0.028	0.017	0.011
14	0.15	0.105	0.072	0.048	0.031	0.02
15	0.215	0.157	0.111	0.077	0.052	0.034
16	0.292	0.221	0.163	0.117	0.082	0.056
17	0.378	0.297	0.227	0.169	0.123	0.087
18	0.469	0.381	0.302	0.233	0.175	0.128
19	0.561	0.47	0.384	0.306	0.238	0.18
20	0.647	0.559	0.471	0.387	0.31	0.243
21	0.726	0.644	0.558	0.472	0.389	0.314
22	0.793	0.721	0.64	0.556	0.472	0.392
23	0.849	0.788	0.716	0.637	0.555	0.473
24	0.893	0.843	0.782	0.712	0.635	0.554
25	0.927	0.888	0.838	0.777	0.708	0.632
26	0.951	0.922	0.883	0.832	0.772	0.704
27	0.969	0.948	0.917	0.878	0.827	0.768
28	0.98	0.966	0.944	0.913	0.873	0.823
29	0.988	0.978	0.963	0.94	0.908	0.868
30	0.993	0.987	0.976	0.96	0.936	0.904
31	0.996	0.992	0.985	0.973	0.956	0.932

Table C (continued)

r	$c' = 19$	$c' = 20$	$c' = 21$	$c' = 22$	$c' = 23$	$c' = 24$
32	0.998	0.995	0.991	0.983	0.971	0.953
33	0.999	0.997	0.994	0.99	0.981	0.969
34	0.999	0.999	0.997	0.994	0.988	0.979
35	1	0.999	0.998	0.996	0.993	0.987
36	1	1	0.999	0.998	0.996	0.992
37	1	1	0.999	0.999	0.997	0.995
38	1	1	1	0.999	0.999	0.997
39	1	1	1	1	0.999	0.998
40	1	1	1	1	1	0.999
41	1	1	1	1	1	0.999
42	1	1	1	1	1	1

r	$c' = 25$	$c' = 26$	$c' = 27$
9	0	0	0
10	0.001	0	0
11	0.001	0.001	0
12	0.003	0.002	0.001
13	0.006	0.004	0.002
14	0.012	0.008	0.005
15	0.022	0.014	0.009
16	0.038	0.025	0.016
17	0.06	0.041	0.027
18	0.092	0.065	0.044
19	0.134	0.097	0.069
20	0.185	0.139	0.101
21	0.247	0.19	0.144
22	0.318	0.252	0.195

Table C (continued)

r	$c' = 25$	$c' = 26$	$c' = 27$
23	0.394	0.321	0.256
24	0.473	0.396	0.324
25	0.553	0.474	0.398
26	0.629	0.552	0.474
27	0.7	0.627	0.551
28	0.763	0.697	0.625
29	0.818	0.759	0.693
30	0.863	0.813	0.755
31	0.9	0.859	0.809
32	0.929	0.896	0.855
33	0.95	0.925	0.892
34	0.966	0.947	0.921
35	0.978	0.964	0.944
36	0.985	0.976	0.961
37	0.991	0.984	0.974
38	0.994	0.99	0.983
39	0.997	0.994	0.989
40	0.998	0.996	0.993
41	0.999	0.998	0.996
42	0.999	0.999	0.997
43	1	0.999	0.998
44	1	1	0.999
45	1	1	0.999
46	1	1	0.999

Appendix B

Basic Programs

This appendix contains BASIC programs which solve the problems in this text and generate data or diagrams. They are written in Microsoft Basic d.

1. Program CNTL CHTS 2

This program calculates the following for samples of ten or less:

a. Natural process limits.
b. Control chart limits.
c. Percentage of production between any two limits.
d. Sample size for given confidence level of detecting a given mean shift.
e. Confidence of detecting a given mean shift with a given sample size.

The accuracy of the integration of the area under the normal curve can be increased by reducing the value of DX in line 1100. This is the width of the trapezoidal slice used in the integration.

```
10 PRINT"FOR SAMPLE SIZE FROM 2 TO 10"
20 PRINT"GIVEN SAMPLE OR POPULATION DATA THIS PROGRAM WILL
CALCULATE"
30 PRINT"1.  PROCESS 3 SIGMA LIMITS"
40 PRINT"2.  CONTROL CHART LIMITS FOR X BAR, R, AND SIGMA CHARTS"
50 PRINT"3.  PERCENT OF PRODUCTION BETWEEN LIMITS"
60 PRINT     "     FOR NORMALLY DISTRIBUTED PROCESS"
70 GOSUB 7000 : REM READ IN  SAMPLE FACTORS
80 PRINT"WHAT IS X DOUBLE BAR OR X BAR PRIME ";
90 INPUT XB:PRINT
100 PRINT" DATA AVAILABLE"
110 PRINT"     1  SIGMA PRIME"
120 PRINT"     2  SIGMA BAR"
130 PRINT"     3  R BAR"
140 INPUT"TYPE NUMBER OF DATA AVAILABLE";DA
150 IF DA >1 THEN 200
155 REM CALCULATE CENTRAL LINES AND LIMITS
160 INPUT "TYPE VALUE OF SIGMA PRIME";SP
170 INPUT "TYPE NUMBER OF MEASUREMENTS IN SAMPLES TO BE USED IN
CHARTS ";N
180 XCL = XB :XUCL = XB + A(N)*SP :XLCL = XB - A(N)*SP :RCL =
DT(N)*SP:RUCL = D2(N)*SP :RLCL = D1(N)*SP :SCL = C2(N)*SP :SUCL =
B2(N)*SP :SLCL = B1(N)*SP
```

```
181 LPRINT"FOR A MEAN OF ";XB;" , SIGMA PRIME OF ";SP; " AND  SAMPLE
SIZE OF "; N

183 LPRINT:LPRINT"X BAR CHART CENTER LINE IS ";XCL
184 LPRINT"          ULC IS ";XUCL
186 LPRINT"          LCL IS ";XLCL
188 LPRINT:LPRINT"R CHART CENTERLINE IS ";RCL
190 LPRINT"       UCL IS ";RUCL
192 LPRINT"        LCL IS ";RLCL
194 LPRINT:LPRINT"SIGMA CHART CENTERLINE IS ";SCL
196 LPRINT"       UCL IS ";SUCL
198 LPRINT"        LCL IS ";SLCL

199 GOTO 500
200 IF DA >2 THEN 300
201 REM CALCULATE CENTRAL LINES AND LIMITS
210 INPUT "TYPE VALUE OF SIGMA BAR";SB
220 INPUT "TYPE NUMBER OF MEASUREMENTS IN SAMPLES ";N
225 LET SP = SB/C2(N) :REM ESTIMATE SIGMA PRIME
230 XCL = XB :XUCL = XB + A1(N)*SB :XLCL = XB - A1(N)*SB  :SCL = SB :SUCL
= B4(N)*SB :SLCL = B3(N)*SB
281 LPRINT"FOR A MEAN OF ";XB;" , SIGMA BAR OF ";SB; " AND  SAMPLE SIZE
OF ";N
```

```
282 LPRINT:LPRINT"X BAR CHART CENTER LINE IS ";XCL
284 LPRINT"          ULC IS ";XUCL
286 LPRINT"          LCL IS ";XLCL
294 LPRINT:LPRINT"SIGMA CHART CENTERLINE IS ";SCL
296 LPRINT"       UCL IS ";SUCL
298 LPRINT"        LCL IS ";SLCL

299 GOTO 500

300 REM CALCULATE CENTRAL LINES AND LIMITS
310 INPUT "TYPE VALUE OF R BAR";RB
320 INPUT "TYPE NUMBER OF MEASUREMENTS IN SAMPLES ";N
325 LET SP = RB/DT(N) :REM ESTIMATE SIGMA PRIME --DT = LITTLE D2
330 XCL = XB :XUCL = XB + A2(N)*RB :XLCL = XB - A2(N)*RB  :RCL = RB :RUCL
= D4(N)*RB :RLCL = D3(N)*RB
381 LPRINT"FOR A MEAN OF ";XB;" , R BAR OF ";RB; " AND  SAMPLE SIZE OF
";N
382 LPRINT:LPRINT"X BAR CHART CENTER LINE IS ";XCL
384 LPRINT"          ULC IS ";XUCL
386 LPRINT"          LCL IS ";XLCL
388 LPRINT:LPRINT"R CHART CENTERLINE IS ";RCL
390 LPRINT"       UCL IS ";RUCL
```

```
392 LPRINT"        LCL IS ";RLCL




500 REM CALCULATE PROCESS LIMITS

505 LET UPL = INT((XB + 3 * SP) *10000 +.5) / 10000

515 LET LPL = INT((XB - 3 * SP) *10000 +.5) / 10000

520 PRINT: PRINT"LOWER PROCESS LIMIT IS ";LPL

525 LPRINT: LPRINT"LOWER PROCESS LIMIT IS ";LPL

530 LPRINT"UPPER PROCESS LIMIT IS ";UPL

535 PRINT"UPPER PROCESS LIMIT IS ";UPL

540 PRINT "SIGMA PRIME IS  "; SP

545 LPRINT "SIGMA PRIME IS  "; SP

600 REM CALCULATE AREA UNDER THE NORMAL CURVE

610 PRINT:PRINT"TO CALCULATE PERCENTAGE OF PRODUCTION BETWEEN

LIMITS TYPE IN LOWER LIMIT"

620 PRINT"TYPE 99 TO END  88 to CALCULATE CONFIDENCE LEVELS"

630 INPUT LL

640 IF LL = 99 THEN 9990

642 IF LL = 88 THEN 8000

643 INPUT"TYPE UPPER LIMIT";UL


644 REM CALCULATION OF AREA UNDER NORMAL CURVE
```

```
645 X0 = (LL - XB) / SP

660 XT = (UL - XB ) / SP

700 PRINT:PRINT"THE WIDER THE LIMITS THE LONGER IT TAKES ME TO
CALCULATE"

1000 LET XL = 0: REM RETURN XL TO 0

1100 DX = .1:REM WIDTH OF AREA SLICE , MAKE SMALLER FOR GREATER
ACCURACY

1200 SM=0

1300 LET X = X0: GOSUB 6000

1400 YL=FX

1500 X1=X+DX

1600 FOR X =  X1 TO XT STEP DX

1700 GOSUB 6000

1800 YT=FX

1900 SM = SM + (YL + YT )/ 2 * DX

2000 YL = YT:XL = X

2100 NEXT X

2200 IF XL > XT THEN 2500

2300 DX = XT - XL:X = XT :GOSUB 6000

2400 SM = SM + (YL + FX) / 2 * DX

2500 REM lprint AREA

2501 SM = INT (SM * 10000 + .5) / 10000

2502 IF QQ = 1 THEN 9160 : REM INDICATES CALCULATION OF PROB OF
```

```
DETECTION

2503 LPRINT"PRODUCTION BETWEEN "; LL; " AND ";UL ;" = ";SM * 100;"

PERCENT"

2504 PRINT"PRODUCTION BETWEEN "; LL; " AND ";UL ;" = ";SM * 100;"

PERCENT"

2510 LPRINT:LPRINT:LPRINT:LPRINT


2520 PRINT:PRINT:PRINT:PRINT:PRINT

2530 PRINT"TYPE 99 TO END 88 TO CALCULATE CONFIDENCE LEVLES"

2540 INPUT LL

2550 IF LL = 88 THEN 8000


5100 GOTO 9990


6000 REM FUNCTION

6100 FX = (1 / (2 * 3.14159) ^ .5 )* (2.7182 ^ (- (X^2)/2))

6200 RETURN


7000 REM  READ IN CONTROL CHART DATA


7002 FOR I%= 2 TO 10

7004 READ N(I%), C2(I%), DT(I%), A(I%), A1(I%),A2(I%),B1(I%),B2(I%),

B3(I%), B4(I%), D1(I%), D2(I%), D3(I%), D4(I%)
```

```
7007 NEXT I%

7009 RETURN

7010 REM N,  C2, DT, A, A1, A2, B1, B2, B3, B4, D1, D2, D3, D4


7020 DATA 2, .5642,1.128,2.121,3.760,1.880, 0,1.843, 0,3.267,0 ,3.686,
0,3.267

7030 DATA 3, .7236,1.693,1.732,2.394,1.023,0, 1.858,0, 2.568,0,4.353,
0,2.575

7040 DATA 4, .7979,2.059,1.5,1.88, .729,0,1.808,0,2.266,0,4.698,0,2.282

7050 DATA 5, .8407,2.326,1.342,1.596,
.577,0,1.756,0,2.089,0,4.918,0,2.115

7060 DATA 6, .8686,2.534,1.225,1.41, .483, .026,1.711, .03,1.97,0,5.078,
0,2.004

7070 DATA 7, .8882,2.704,1.134,1.277,.419,.105,1.672,.118,1.882, .205,
5.203,  .076,1.924

7080 DATA 8, .9027,2.847,1.061,1.175, .373,.167,1.638, .185,1.815, .387,
5.307,  .136,1.864

7090 DATA 9, .9139,2.970,1,1.09, .337, .219,1.609, .239,1.761, .546,5.394,
.184,1.816

7100 DATA 10, .9227,3.078, .949,1.028, .308, .262,1.584, .284,1.716,
.687,5.469, .223,1.777


8000 REM CALCULATE SAMPLE SIZE AND CONFIDENCE LEVELS
```

```
8010 PRINT:PRINT

8015 LPRINT:LPRINT

8020 PRINT"THIS SECTION WILL CALCULATE"

8030 PRINT"     1.  SAMPLE SIZE FOR A GIVEN CONFIDENCE LEVEL"

8040 PRINT"     2.  CONFIDENCE LEVEL OF DETECTING GIVEN MEAN SHIFT IN
1st SAMPLE"

8050 PRINT:PRINT"TYPE NUMBER OF YOUR CHOICE, 99 TO QUIT";

8060 INPUT LL

8070 IF LL = 99 THEN 9990

8090 IF LL = 2 THEN 9000


8100 REM THIS SECTION WILL CALCULATE SAMPLE SIZE

8110 PRINT:PRINT

8120 PRINT"WHAT NEW MEAN VALUE DO YOU WISH TO DETECT";

8130 INPUT XS

8140 PRINT:PRINT"FOR 80% LEVEL Z = .84, 85% Z = 1.04, 90% Z = 1.29, 95%
Z = 1.65, 99% Z = 2.33"

8150 PRINT"TYPE IN DESIRED Z VALUE"; : INPUT Z

8170 LET N = (SP*(Z+3)/(XCL-XS))^2

8180 PRINT:PRINT"FOR NEW MEAN OF ";XS;" AND Z = ";Z;" SAMPLE SIZE IS ";N

8190 LPRINT:LPRINT"FOR NEW MEAN OF ";XS;" AND Z = ";Z;" SAMPLE SIZE IS
";N
```

```
8200 PRINT:PRINT "TYPE 17 TO CALCULATE NEW CHART VALUES, 99 TO
QUIT"
8210 INPUT LL
8230 IF LL = 99 THEN 9990
8240 GOTO 170

9000 REM CALCULATION OF CONFIDENCE LEVEL OF DETECTION
9002 LET SP = SP / (N)^.5 : REM CHANGES SP TO SIGMA X BAR
9005 PRINT"WHAT NEW MEAN VALUE DO YOU WISH TO DETECT";
9007 INPUT XS
9010 IF XS < XCL THEN 9050: REM MEANS SHIFT BELOW CENTERLINE
9020 REM CALC AREA ABOVE UCL
9030 LET UL = XS  + 3 * SP
9040 LET LL = XUCL
9045 GOTO 9080
9050 REM CALC AREA BELOW LCL
9060 LET UL = XLCL
9070 LET LL = XS - 3 * SP
9080 LET XB = XS
9150 LET QQ = 1
9151 GOTO 645: REM GET AREA UNDER CURVE
9160 PRINT:PRINT"THE PROBABILITY OF DETECTING MEAN SHIFT TO ";XS
9170 PRINT"WITH SAMPLE SIZE ";N;" IS "; SM * 100;" PERCENT"
```

```
9180 LPRINT:LPRINT"THE PROBABILITY OF DETECTING MEAN SHIFT TO ";XS

9190 LPRINT"WITH SAMPLE SIZE ";N;" IS "; SM * 100;" PERCENT"



9990

LPRINT"________________________________________________________"

9991 LPRINT:LPRINT:LPRINT:LPRINT:PRINT:PRINT:PRINT:PRINT

9992 END
```

2. Program Area 3 + 3

This program draws the normal curve with an area between limits shaded. The limits of the shaded area are in line 1360.

```
10 FOR I= 1 TO 15 : PRINT : NEXT I

100 CLS

1200 FOR X = -3.5 TO 3 STEP .05

1210 X = INT (X* 100 +.5)/100

1290 FX = (1/(2*3.1416)^.5)*(2.718 ^(-(X^2)/2))

1295 Y = (X+3.5)*60

1300 FX=FX*350

1351 PRESET (Y, 150-FX), 33

1360 IF (X>= -3) AND (X<=3)THEN GOSUB 2000

1370 IF X = INT(X) THEN GOSUB 3000

1380 IF X = INT (X) THEN GOSUB 4000

1400 NEXT X

1500 LINE(0,150)-(400,150)

1999 GOTO 9999

2000 REM    SHADE

2100 LINE (Y, 150)-(Y,  150-FX)

2200 RETURN
```

```
2999 GOTO 9999

3000 REM BOTTOM LINE

3005 IF Q = > 1 THEN 3045

3010 FOR I= 1 TO 10 : PRINT : NEXT I

3015 Q=Q+1

3045 PRINT PTAB(Y-10); INT(X);

3050 RETURN

3999 GOTO 9999

4000 REM TIC MARKS

4100 LINE (Y, 150)-(Y, 160)

4500 RETURN

9999 END
```

3. Program BINOMIAL

This program calculates the probability of exactly r and of r or less for the binomial distribution.

```
10 PRINT"THIS PROGRAM CALCULATES THE PROBABILITY OF R OR LESS
OCCURANCES"
12 PRINT"OF AN EVENT WITH A CONSTANT PROBABITITY OF P'"
14 PRINT"IN N TRIALS"
20 REM THIS SECTION WILL GET IN VALUES FOR N, P', AND THE MAX VALUE
OF R"
30 PRINT"HOW MANY TRIALS";
40 INPUT N
50 PRINT"TYPE IN THE VALUE OF R";
60 INPUT RMAX
70 PRINT
75 DIM C(RMAX): DIM BCUM(RMAX):DIM BP(RMAX)
80 PRINT"WHAT IS P'";
90 INPUT P
95 REM FOR COMBINATIONS THE FORMULA IS C=N!/(R!(N-R)!), RF= R!, QF =
N!/(N-R)!
100 REM THIS SECTION WILL CALCULATE A C() VALUE FOR R = 0 TO RMAX
105 FOR R = 0 TO RMAX
```

```
120 LET Z = R

130 GOSUB 6000

140 LET RF = ZF

148 IF N =>33 THEN 500

150 LET Z = N

160 GOSUB 6000

170 LET NF = ZF

180 LET NR = N - R

190 LET Z = NR

200 GOSUB 6000

210 LET DF = ZF

400 REM THIS SECTION CALCULATES C FOR N <33

410 LET C = NF/RF/DF

420 GOTO 530

500 REM THIS SECTION CALCULATES C FOR N >= 33

510 GOSUB 4000

520 LET C = QF/RF

530 LET C(R)  = C

540 LET C = 0

550 NEXT R


600 REM THIS SECTION CALCULATES THE BINOMIAL PROBABILITY, BP, FOR
EACH VALUE OF C()
```

```
610 FOR R = 0 TO RMAX

620 LET BP(R) = C(R) * (1 - P)^(N - R) * P ^ R

625 LET BCUM = BCUM + BP(R)

627 LET BCUM(R) = BCUM

630 NEXT R

700 CLS

705 LPRINT" FOR N = ";N;" P' = ";P;" R = ";R

710 PRINT"   BINOMIAL PROBABILITIES"

711 LPRINT"   BINOMIAL PROBABILITIES"

720 LPRINT

721 PRINT

730 PRINT"R","PROB OF EXACTLY R", "CUMULATIVE PROBABILITY"

740 PRINT"-","-----------------","----------------------"

750 PRINT

751 LPRINT"R","PROB OF EXACTLY R", "CUMULATIVE PROBABILITY"

752 LPRINT"-","-----------------","----------------------"

753 LPRINT


760 FOR R = 0 TO RMAX

765 LET BP(R) = INT(BP(R) * 100000! + .5 )/ 100000!

767 LET BCUM(R) = INT(BCUM(R) * 100000! + .5 )/ 100000!

770 PRINT R,BP(R),,BCUM(R)

770 LPRINT R,BP(R),,BCUM(R)
```

```
780 NEXT R

790 PRINT


3999 GOTO 9000

4000 REM THIS SECTION CALCULATES N!/(N-R)! AND CALS IT QF

4010 LET QF = N

4015 IF R = 1 THEN 4200

4017 IF R = 0 THEN 4300

4020 FOR Q = 1 TO R-1

LET QF = QF * (N-Q)

4050 NEXT Q

4100 GOTO 4500

4200 LET QF = N

4250 GOTO 4500

4300 LET QF = 1

4500 RETURN

5999 GOTO 9000

6000 REM THIS SECTION CALCULATES Z! CALLED ZF

6010 IF Z = 0 THEN 6900

6020 IF Z = 1 THEN 6900

6030 LET ZF = Z

6100 FOR ZZ = 1 TO Z - 1

6110 LET ZF = ZF * (Z - ZZ)
```

```
6120 NEXT ZZ

6800 GOTO 7000

6900 LET ZF = 1

7000 RETURN


9000 REM THIS SECTION WRITES ON CLIPBOARD FILE

9002 PRINT"TYPE 1 TO WRITE ON CLIP, 2 TO QUIT"

9004 INPUT CH

9006 IF CH = 2 THEN 9999

9008 PRINT"WRITING"

9010 OPEN "CLIP:" FOR OUTPUT AS #1

9020 WRITE #1, N,P,0

9760 FOR R = 0 TO RMAX

9770 WRITE #1,R,BP(R),BCUM(R)

9780 NEXT R

9800 CLOSE #1

9999 END
```

4. Program NORMAL1.111L

This program generates a list of 100 numbers with a normal distribution of the decimal values and it counts the frequency. The mean and standard deviation of the decimal values are in lines 20 and 30. The base value is in lines 170 and 220.

```
10 REM M= MEAN    SI = STD DEV

20 LET M=100

30 SI=25

35 DIM C2(100)

40 DIM C(1000)

41 DIM F(102)

50 FOR J% = 1 TO 100 : REM GENERATE 100 NUMBERS

60 REM LINES 70 TO 110 GENERATE INTEGER

70 FOR I% = 1 TO 12

80 LET S = S + RND (1): REM  SAME SEQUENCE

90 NEXT I%

100 F% = M+SI * (S-6)

110 LET F(J%)  = F%

120 S=0

130 C(F%)=C(F%) +1

140 NEXT J%
```

```
141 FOR I%= 1 TO 100

142 LET N% = INT(F(I%) +.5)/10

144 LET C2(N%) = C2(N%) + 1

147 NEXT I%

150 FOR I% = 0 TO 25

154 NEXT I%

155 OPEN "CLIP:" FOR OUTPUT AS #1

160 FOR F% = 1 TO 20

170 WRITE #1, F%/100 +32,C2(F%)

200 NEXT F%

210 FOR I=1 TO 100

220 LET F(I) = F(I)/1000 + 32

230 NEXT I

300 FOR I =1 TO 102 STEP 6

310 WRITE #1, F(I),F(I+1),F(I+2),F(I+3),F(I+4),F(I+5)

350 NEXT I

500 CLOSE #1

999 END
```

5. Program POISSON

This program calculates the probability or exactly r and of r or less using the Poisson probability.

```
10 PRINT"THIS PROGRAM CALCULATES THE PROBABILITY OF R OR LESS
OCCURANCES"
12 PRINT"OF AN EVENT WITH AN AVERAGE OCCURRANCE OF C' "
20 REM THIS SECTION WILL GET IN VALUES FOR C' AND THE MAX VALUE OF R"
50 PRINT"TYPE IN THE VALUE OF R";
60 INPUT RMAX
65 DIM FR(RMAX):DIM PB(RMAX)
70 PRINT
80 PRINT"WHAT IS C'";
90 INPUT C

100 REM THIS SECTION WILL CALCULATE A C() VALUE FOR R = 0 TO RMAX
105 FOR R = 0 TO RMAX
120 LET Z = R
130 GOSUB 6000

530 LET FR(R)  = ZF
550 NEXT R
```

```
600 REM THIS SECTION CALCULATES THE POISSON PROBABILITY,PB, FOR
EACH VALUE OF R
610 FOR R = 0 TO RMAX
620 LET PB(R) = C^R * 2.716^(-C) / FR(R)
625 LET BCUM = BCUM + PB(R)
630 NEXT R
700 CLS
705 LPRINT" FOR C' = ";C;
710 PRINT"   POISSON PROBABILITIES"
711 LPRINT"   POISSON PROBABILITIES"
720 LPRINT
721 PRINT
730 PRINT"R","PROB OF EXACTLY R"
740 PRINT"-","-----------------"
750 PRINT
751 LPRINT"R","PROB OF EXACTLY R"
752 LPRINT"-","-----------------"
753 LPRINT
```

```
760 FOR R = 0 TO RMAX

770 PRINT R,PB(R)

770 LPRINT R,PB(R)

780 NEXT R

790 PRINT

795 PRINT "PROBABILITY OF  "; RMAX; " OR LESS IS"

798 PRINT BCUM

800 LPRINT "PROBABILITY OF  "; RMAX; " OR LESS IS"

801 LPRINT BCUM


3999 GOTO 9999

6000 REM THIS SECTION CALCULATES Z! CALLED ZF

6010 IF Z = 0 THEN 6900

6020 IF Z = 1 THEN 6900

6030 LET ZF = Z

6100 FOR ZZ = 1 TO Z - 1

6110 LET ZF = ZF * (Z - ZZ)

6120 NEXT ZZ

6800 GOTO 7000

6900 LET ZF = 1

7000 RETURN

9999 END
```

6. Program SAM 20

This program produces twenty samples of five values. The mean and standard deviation are in lines 20 and 30. The number of samples and the size of the samples are controlled in lines 50, 142, 144, 300 and 310.

```
5 CLS

10 REM M= MEAN    SI = STD DEV

20 LET M=12

30 SI=2

35 DIM C2(100)

40 DIM C(1000)

41 DIM F(500)

50 FOR J% = 1 TO 100

60 REM LINES 70 TO 110 GENERATE INTEGER

70 FOR I% = 1 TO 12

80 LET S = S + RND (1): REM  SAME SEQUENCE

90 NEXT I%

100 F = M+SI * (S-6)

110 LET F(J%) = INT(F * 100+.5)/100

120 S=0

140 NEXT J%

142 FOR I=1 TO 100 STEP 5
```

```
144 PRINT  F(I),F(I+1),F(I+2),F(I+3),F(I+4)

148 NEXT I

155 OPEN "CLIP:" FOR OUTPUT AS #1

300 FOR I =1 TO 100 STEP 5

310 WRITE #1, F(I),F(I+1),F(I+2),F(I+3),F(I+4)

350 NEXT I

500 CLOSE #1

999 END
```

7. Program STD CALC

This program calculates the standard deviation of 100 numbers in the data lines. For other than 100 numbers, lines 5, 10, 70, 100 and 210 must be changed.

```
5 DIM X(100)

10 FOR I =  1 TO 100

40 READ X(I)

50 NEXT I

60 REM SUM

70 FOR I = 1 TO 100

80 LET SUM = SUM + X(I)

90 NEXT I

100 LET XB = SUM /100

110 LET SUM = 0

200 REM ADD[ X(I) - XB ]SQD

210 FOR I=1 TO 100

220 LET SUM =  SUM + (X(I) - XB)^2

230 NEXT I

240 LET SG = (SUM/100)^.5

300 REM PRINT RESULT

310 PRINT"STD DEV = ";SG
```

```
320 PRINT

330 PRINT"MEAN = ";XB

350 LIST

1000 DATA 1.061,1.111,1.116,1.085,1.097,1.1

1010 DATA 1.083,1.052,1.103,1.118,1.09,1.072

1020 DATA 1.1,1.092,1.07,1.133,1.134,1.075

1030 DATA 1.107,1.102,1.127,1.121,1.086,1.093

1040 DATA 1.125,1.105,1.115,1.097,1.15,1.107

1050 DATA 1.079,1.104,1.049,1.129,1.111,1.104

1060 DATA 1.142,1.099,1.095,1.088,1.076,1.095

1070 DATA 1.091,1.07,1.098,1.118,1.132,1.142

1080 DATA 1.129,1.073,1.102,1.093,1.141,1.136

1090 DATA 1.111,1.063,1.13,1.136,1.116,1.118

1100 DATA 1.092,1.074,1.104,1.147,1.125,1.08

1110 DATA 1.106,1.067,1.074,1.08,1.083,1.067

1120 DATA 1.111,1.128,1.142,1.063,1.108,1.126

1130 DATA 1.144,1.12,1.136,1.127,1.047,1.128

1140 DATA 1.092,1.133,1.111,1.102,1.097,1.147

1150 DATA 1.068,1.108,1.097,1.073,1.079,1.085

1160 DATA 1.139,1.137,1.077,1.074
```

Appendix
C

Standard Sampling Plans

MIL-STD-105D

TABLE I—Sample size code letters

(See 9.2 and 9.3)

Lot or batch size			Special inspection levels				General inspection levels		
			S-1	S-2	S-3	S-4	I	II	III
2	to	8	A	A	A	A	A	A	B
9	to	15	A	A	A	A	A	B	C
16	to	25	A	A	B	B	B	C	D
26	to	50	A	B	B	C	C	D	E
51	to	90	B	B	C	C	C	E	F
91	to	150	B	B	C	D	D	F	G
151	to	280	B	C	D	E	E	G	H
281	to	500	B	C	D	E	F	H	J
501	to	1200	C	C	E	F	G	J	K
1201	to	3200	C	D	E	G	H	K	L
3201	to	10000	C	D	F	G	J	L	M
10001	to	35000	C	D	F	H	K	M	N
35001	to	150000	D	E	G	J	L	N	P
150001	to	500000	D	E	G	J	M	P	Q
500001	and	over	D	E	H	K	N	Q	R

SINGLE
NORMAL

TABLE II-A—Single sampling plans for normal inspection (Master table)

(See 9.4 and 9.5)

Sample size code letter	Sample size	Acceptable Quality Levels (normal inspection)																									
		0.010	0.015	0.025	0.040	0.065	0.10	0.15	0.25	0.40	0.65	1.0	1.5	2.5	4.0	6.5	10	15	25	40	65	100	150	250	400	650	1000
		Ac Re	Ac Re	Ac Re	Ac Re	Ac Re	Ac Re	Ac Re	Ac Re	Ac Re	Ac Re	Ac Re	Ac Re	Ac Re	Ac Re	Ac Re	Ac Re	Ac Re	Ac Re	Ac Re	Ac Re	Ac Re	Ac Re	Ac Re	Ac Re	Ac Re	Ac Re
A	2	↓	↓	↓	↓	↓	↓	↓	↓	↓	↓	↓	↓	↓	↓	0 1	↓	↓	1 2	2 3	3 4	5 6	7 8	10 11	14 15	21 22	30 31
B	3	↓	↓	↓	↓	↓	↓	↓	↓	↓	↓	↓	↓	↓	0 1	↑	↓	1 2	2 3	3 4	5 6	7 8	10 11	14 15	21 22	30 31	44 45
C	5	↓	↓	↓	↓	↓	↓	↓	↓	↓	↓	↓	↓	0 1	↑	↓	1 2	2 3	3 4	5 6	7 8	10 11	14 15	21 22	30 31	44 45	↑
D	8	↓	↓	↓	↓	↓	↓	↓	↓	↓	↓	↓	0 1	↑	↓	1 2	2 3	3 4	5 6	7 8	10 11	14 15	21 22	30 31	44 45	↑	↑
E	13	↓	↓	↓	↓	↓	↓	↓	↓	↓	↓	0 1	↑	↓	1 2	2 3	3 4	5 6	7 8	10 11	14 15	21 22	30 31	44 45	↑	↑	↑
F	20	↓	↓	↓	↓	↓	↓	↓	↓	↓	0 1	↑	↓	1 2	2 3	3 4	5 6	7 8	10 11	14 15	21 22	↑	↑	↑	↑	↑	↑
G	32	↓	↓	↓	↓	↓	↓	↓	↓	0 1	↑	↓	1 2	2 3	3 4	5 6	7 8	10 11	14 15	21 22	↑	↑	↑	↑	↑	↑	↑
H	50	↓	↓	↓	↓	↓	↓	↓	0 1	↑	↓	1 2	2 3	3 4	5 6	7 8	10 11	14 15	21 22	↑	↑	↑	↑	↑	↑	↑	↑
J	80	↓	↓	↓	↓	↓	↓	0 1	↑	↓	1 2	2 3	3 4	5 6	7 8	10 11	14 15	21 22	↑	↑	↑	↑	↑	↑	↑	↑	↑
K	125	↓	↓	↓	↓	↓	0 1	↑	↓	1 2	2 3	3 4	5 6	7 8	10 11	14 15	21 22	↑	↑	↑	↑	↑	↑	↑	↑	↑	↑
L	200	↓	↓	↓	↓	0 1	↑	↓	1 2	2 3	3 4	5 6	7 8	10 11	14 15	21 22	↑	↑	↑	↑	↑	↑	↑	↑	↑	↑	↑
M	315	↓	↓	↓	0 1	↑	↓	1 2	2 3	3 4	5 6	7 8	10 11	14 15	21 22	↑	↑	↑	↑	↑	↑	↑	↑	↑	↑	↑	↑
N	500	↓	↓	0 1	↑	↓	1 2	2 3	3 4	5 6	7 8	10 11	14 15	21 22	↑	↑	↑	↑	↑	↑	↑	↑	↑	↑	↑	↑	↑
P	800	↓	0 1	↑	↓	1 2	2 3	3 4	5 6	7 8	10 11	14 15	21 22	↑	↑	↑	↑	↑	↑	↑	↑	↑	↑	↑	↑	↑	↑
Q	1250	0 1	↑	↓	1 2	2 3	3 4	5 6	7 8	10 11	14 15	21 22	↑	↑	↑	↑	↑	↑	↑	↑	↑	↑	↑	↑	↑	↑	↑
R	2000	↑	↑	1 2	2 3	3 4	5 6	7 8	10 11	14 15	21 22	↑	↑	↑	↑	↑	↑	↑	↑	↑	↑	↑	↑	↑	↑	↑	↑

↓ = Use first sampling plan below arrow. If sample size equals, or exceeds, lot or batch size, do 100 percent inspection.
↑ = Use first sampling plan above arrow.
Ac = Acceptance number.
Re = Rejection number.

TABLE II-B — Single sampling plans for tightened inspection (Master table)

(See 9.4 and 9.5)

Acceptable Quality Levels (tightened inspection) — each cell gives Ac Re

Sample size code letter	Sample size	0.010	0.015	0.025	0.040	0.065	0.10	0.15	0.25	0.40	0.65	1.0	1.5	2.5	4.0	6.5	10	15	25	40	65	100	150	250	400	650	1000
		Ac Re	Ac Re	Ac Re	Ac Re	Ac Re	Ac Re	Ac Re	Ac Re	Ac Re	Ac Re	Ac Re	Ac Re	Ac Re	Ac Re	Ac Re	Ac Re	Ac Re	Ac Re	Ac Re	Ac Re	Ac Re	Ac Re	Ac Re	Ac Re	Ac Re	Ac Re
A	2	↓	↓	↓	↓	↓	↓	↓	↓	↓	↓	↓	↓	↓	↓	↓	↓	↓	↓	1 2	2 3	3 4	5 6	8 9	12 13	18 19	27 28
B	3	↓	↓	↓	↓	↓	↓	↓	↓	↓	↓	↓	↓	↓	↓	0 1	↓	↓	1 2	2 3	3 4	5 6	8 9	12 13	18 19	27 28	41 42
C	5	↓	↓	↓	↓	↓	↓	↓	↓	↓	↓	↓	↓	↓	0 1	↓	↓	1 2	2 3	3 4	5 6	8 9	12 13	18 19	27 28	41 42	↑
D	8	↓	↓	↓	↓	↓	↓	↓	↓	↓	↓	↓	↓	0 1	↓	↓	1 2	2 3	3 4	5 6	8 9	12 13	18 19	27 28	41 42	↑	↑
E	13	↓	↓	↓	↓	↓	↓	↓	↓	↓	↓	↓	0 1	↓	↓	1 2	2 3	3 4	5 6	8 9	12 13	18 19	27 28	41 42	↑	↑	↑
F	20	↓	↓	↓	↓	↓	↓	↓	↓	↓	↓	0 1	↓	↓	1 2	2 3	3 4	5 6	8 9	12 13	18 19	↑	↑	↑	↑	↑	↑
G	32	↓	↓	↓	↓	↓	↓	↓	↓	↓	0 1	↓	↓	1 2	2 3	3 4	5 6	8 9	12 13	18 19	↑	↑	↑	↑	↑	↑	↑
H	50	↓	↓	↓	↓	↓	↓	↓	↓	0 1	↓	↓	1 2	2 3	3 4	5 6	8 9	12 13	18 19	↑	↑	↑	↑	↑	↑	↑	↑
J	80	↓	↓	↓	↓	↓	↓	↓	0 1	↓	↓	1 2	2 3	3 4	5 6	8 9	12 13	18 19	↑	↑	↑	↑	↑	↑	↑	↑	↑
K	125	↓	↓	↓	↓	↓	↓	0 1	↓	↓	1 2	2 3	3 4	5 6	8 9	12 13	18 19	↑	↑	↑	↑	↑	↑	↑	↑	↑	↑
L	200	↓	↓	↓	↓	↓	0 1	↓	↓	1 2	2 3	3 4	5 6	8 9	12 13	18 19	↑	↑	↑	↑	↑	↑	↑	↑	↑	↑	↑
M	315	↓	↓	↓	↓	0 1	↓	↓	1 2	2 3	3 4	5 6	8 9	12 13	18 19	↑	↑	↑	↑	↑	↑	↑	↑	↑	↑	↑	↑
N	500	↓	↓	↓	0 1	↓	↓	1 2	2 3	3 4	5 6	8 9	12 13	18 19	↑	↑	↑	↑	↑	↑	↑	↑	↑	↑	↑	↑	↑
P	800	↓	↓	0 1	↓	↓	1 2	2 3	3 4	5 6	8 9	12 13	18 19	↑	↑	↑	↑	↑	↑	↑	↑	↑	↑	↑	↑	↑	↑
Q	1250	↓	0 1	↓	↓	1 2	2 3	3 4	5 6	8 9	12 13	18 19	↑	↑	↑	↑	↑	↑	↑	↑	↑	↑	↑	↑	↑	↑	↑
R	2000	0 1	↑	↓	1 2	2 3	3 4	5 6	8 9	12 13	18 19	↑	↑	↑	↑	↑	↑	↑	↑	↑	↑	↑	↑	↑	↑	↑	↑
S	3150			1 2																							

↓ = Use first sampling plan below arrow. If sample size equals or exceeds lot or batch size, do 100 percent inspection.
↑ = Use first sampling plan above arrow.
Ac = Acceptance number.
Re = Rejection number.

SINGLE
TIGHTENED

SINGLE
REDUCED

TABLE II-C — Single sampling plans for reduced inspection (Master table)

(See 9.4 and 9.5)

Sample size code letter	Sample size	Acceptable Quality Levels (reduced inspection)†																									
		0.010	0.015	0.025	0.040	0.065	0.10	0.15	0.25	0.40	0.65	1.0	1.5	2.5	4.0	6.5	10	15	25	40	65	100	150	250	400	650	1000
		Ac Re	Ac Re	Ac Re	Ac Re	Ac Re	Ac Re	Ac Re	Ac Re	Ac Re	Ac Re	Ac Re	Ac Re	Ac Re	Ac Re	Ac Re	Ac Re	Ac Re	Ac Re	Ac Re	Ac Re	Ac Re	Ac Re	Ac Re	Ac Re	Ac Re	Ac Re
A	2	↓	↓	↓	↓	↓	↓	↓	↓	↓	↓	↓	↓	↓	↓	0 1	↓	↓	1 2	2 3	3 4	5 6	7 8	10 11	14 15	21 22	30 31
B	2	↓	↓	↓	↓	↓	↓	↓	↓	↓	↓	↓	↓	↓	0 1	↑	↓	0 2	1 3	2 4	3 5	5 6	7 8	10 11	14 15	21 22	30 31
C	2	↓	↓	↓	↓	↓	↓	↓	↓	↓	↓	↓	↓	0 1	↑	↓	0 2	1 3	1 4	2 5	3 6	5 8	7 10	10 13	14 17	21 24	↑
D	3	↓	↓	↓	↓	↓	↓	↓	↓	↓	↓	↓	0 1	↑	↓	0 2	1 3	1 4	2 5	3 6	5 8	7 10	10 13	14 17	21 24	↑	↑
E	5	↓	↓	↓	↓	↓	↓	↓	↓	↓	↓	0 1	↑	↓	0 2	1 3	1 4	2 5	3 6	5 8	7 10	10 13	14 17	21 24	↑	↑	↑
F	8	↓	↓	↓	↓	↓	↓	↓	↓	↓	0 1	↑	↓	0 2	1 3	1 4	2 5	3 6	5 8	7 10	10 13	↑	↑	↑	↑	↑	↑
G	13	↓	↓	↓	↓	↓	↓	↓	↓	0 1	↑	↓	0 2	1 3	1 4	2 5	3 6	5 8	7 10	10 13	↑	↑	↑	↑	↑	↑	↑
H	20	↓	↓	↓	↓	↓	↓	↓	0 1	↑	↓	0 2	1 3	1 4	2 5	3 6	5 8	7 10	10 13	↑	↑	↑	↑	↑	↑	↑	↑
J	32	↓	↓	↓	↓	↓	↓	0 1	↑	↓	0 2	1 3	1 4	2 5	3 6	5 8	7 10	10 13	↑	↑	↑	↑	↑	↑	↑	↑	↑
K	50	↓	↓	↓	↓	↓	0 1	↑	↓	0 2	1 3	1 4	2 5	3 6	5 8	7 10	10 13	↑	↑	↑	↑	↑	↑	↑	↑	↑	↑
L	80	↓	↓	↓	↓	0 1	↑	↓	0 2	1 3	1 4	2 5	3 6	5 8	7 10	10 13	↑	↑	↑	↑	↑	↑	↑	↑	↑	↑	↑
M	125	↓	↓	↓	0 1	↑	↓	0 2	1 3	1 4	2 5	3 6	5 8	7 10	10 13	↑	↑	↑	↑	↑	↑	↑	↑	↑	↑	↑	↑
N	200	↓	↓	0 1	↑	↓	0 2	1 3	1 4	2 5	3 6	5 8	7 10	10 13	↑	↑	↑	↑	↑	↑	↑	↑	↑	↑	↑	↑	↑
P	315	↓	0 1	↑	↓	0 2	1 3	1 4	2 5	3 6	5 8	7 10	10 13	↑	↑	↑	↑	↑	↑	↑	↑	↑	↑	↑	↑	↑	↑
Q	500	0 1	↑	↓	0 2	1 3	1 4	2 5	3 6	5 8	7 10	10 13	↑	↑	↑	↑	↑	↑	↑	↑	↑	↑	↑	↑	↑	↑	↑
R	800	↑	↑	0 2	1 3	1 4	2 5	3 6	5 8	7 10	10 13	↑	↑	↑	↑	↑	↑	↑	↑	↑	↑	↑	↑	↑	↑	↑	↑

↓ = Use first sampling plan below arrow. If sample size equals or exceeds lot or batch size, do 100 percent inspection.
↑ = Use first sampling plan above arrow.
Ac = Acceptance number.
Re = Rejection number.
† = If the acceptance number has been exceeded, but the rejection number has not been reached, accept the lot, but reinstate normal inspection (see 10.1.4).

TABLE III-A—Double sampling plans for normal inspection (Master table)

(See 9.4 and 9.5)

Sample size code letter	Sample	Sample size	Cumulative sample size	Acceptable Quality Levels (normal inspection)																									
				0.010	0.015	0.025	0.040	0.065	0.10	0.15	0.25	0.40	0.65	1.0	1.5	2.5	4.0	6.5	10	15	25	40	65	100	150	250	400	650	1000
				Ac Re	Ac Re	Ac Re	Ac Re	Ac Re	Ac Re	Ac Re	Ac Re	Ac Re	Ac Re	Ac Re	Ac Re	Ac Re	Ac Re	Ac Re	Ac Re	Ac Re	Ac Re	Ac Re	Ac Re	Ac Re	Ac Re	Ac Re	Ac Re	Ac Re	Ac Re
A				↓	↓	↓	↓	↓	↓	↓	↓	↓	↓	↓	↓	↓	↓	•	↓	↓	•	•	•	•	•	•	•	•	•
B	First	2	2	↓	↓	↓	↓	↓	↓	↓	↓	↓	↓	↓	↓	↓	•	↑	↓	0 2	0 3	1 4	2 5	3 7	5 9	7 11	11 16	17 22	25 31
	Second	2	4																	1 2	3 4	4 5	6 7	8 9	12 13	18 19	26 27	37 38	56 57
C	First	3	3	↓	↓	↓	↓	↓	↓	↓	↓	↓	↓	↓	↓	•	↑	↓	0 2	0 3	1 4	2 5	3 7	5 9	7 11	11 16	17 22	25 31	↑
	Second	3	6																1 2	3 4	4 5	6 7	8 9	12 13	18 19	26 27	37 38	56 57	
D	First	5	5	↓	↓	↓	↓	↓	↓	↓	↓	↓	↓	↓	•	↑	↓	0 2	0 3	1 4	2 5	3 7	5 9	7 11	11 16	17 22	25 31	↑	↑
	Second	5	10															1 2	3 4	4 5	6 7	8 9	12 13	18 19	26 27	37 38	56 57		
E	First	8	8	↓	↓	↓	↓	↓	↓	↓	↓	↓	↓	•	↑	↓	0 2	0 3	1 4	2 5	3 7	5 9	7 11	11 16	17 22	25 31	↑	↑	↑
	Second	8	16													1 2	3 4	4 5	6 7	8 9	12 13	18 19	26 27	37 38	56 57				
F	First	13	13	↓	↓	↓	↓	↓	↓	↓	↓	↓	•	↑	↓	0 2	0 3	1 4	2 5	3 7	5 9	7 11	11 16	↑	↑	↑	↑	↑	↑
	Second	13	26													1 2	3 4	4 5	6 7	8 9	12 13	18 19	26 27						
G	First	20	20	↓	↓	↓	↓	↓	↓	↓	↓	•	↑	↓	0 2	0 3	1 4	2 5	3 7	5 9	7 11	11 16	↑	↑	↑	↑	↑	↑	↑
	Second	20	40												1 2	3 4	4 5	6 7	8 9	12 13	18 19	26 27							
H	First	32	32	↓	↓	↓	↓	↓	↓	↓	•	↑	↓	0 2	0 3	1 4	2 5	3 7	5 9	7 11	11 16	↑	↑	↑	↑	↑	↑	↑	↑
	Second	32	64											1 2	3 4	4 5	6 7	8 9	12 13	18 19	26 27								
J	First	50	50	↓	↓	↓	↓	↓	↓	•	↑	↓	0 2	0 3	1 4	2 5	3 7	5 9	7 11	11 16	↑	↑	↑	↑	↑	↑	↑	↑	↑
	Second	50	100									1 2	3 4	4 5	6 7	8 9	12 13	18 19	26 27										
K	First	80	80	↓	↓	↓	↓	↓	•	↑	↓	0 2	0 3	1 4	2 5	3 7	5 9	7 11	11 16	↑	↑	↑	↑	↑	↑	↑	↑	↑	↑
	Second	80	160									1 2	3 4	4 5	6 7	8 9	12 13	18 19	26 27										
L	First	125	125	↓	↓	↓	↓	•	↑	↓	0 2	0 3	1 4	2 5	3 7	5 9	7 11	11 16	↑	↑	↑	↑	↑	↑	↑	↑	↑	↑	↑
	Second	125	250								1 2	3 4	4 5	6 7	8 9	12 13	18 19	26 27											
M	First	200	200	↓	↓	↓	•	↑	↓	0 2	0 3	1 4	2 5	3 7	5 9	7 11	11 16	↑	↑	↑	↑	↑	↑	↑	↑	↑	↑	↑	↑
	Second	200	400							1 2	3 4	4 5	6 7	8 9	12 13	18 19	26 27												
N	First	315	315	↓	↓	•	↑	↓	0 2	0 3	1 4	2 5	3 7	5 9	7 11	11 16	↑	↑	↑	↑	↑	↑	↑	↑	↑	↑	↑	↑	↑
	Second	315	630						1 2	3 4	4 5	6 7	8 9	12 13	18 19	26 27													
P	First	500	500	↓	•	↑	↓	0 2	0 3	1 4	2 5	3 7	5 9	7 11	11 16	↑	↑	↑	↑	↑	↑	↑	↑	↑	↑	↑	↑	↑	↑
	Second	500	1000					1 2	3 4	4 5	6 7	8 9	12 13	18 19	26 27														
Q	First	800	800	•	↑	↓	0 2	0 3	1 4	2 5	3 7	5 9	7 11	11 16	↑	↑	↑	↑	↑	↑	↑	↑	↑	↑	↑	↑	↑	↑	↑
	Second	800	1600				1 2	3 4	4 5	6 7	8 9	12 13	18 19	26 27															
R	First	1250	1250	↑	↑	0 2	0 3	1 4	2 5	3 7	5 9	7 11	11 16	↑	↑	↑	↑	↑	↑	↑	↑	↑	↑	↑	↑	↑	↑	↑	↑
	Second	1250	2500			1 2	3 4	4 5	6 7	8 9	12 13	18 19	26 27																

↓ = Use first sampling plan below arrow. If sample size equals or exceeds lot or batch size, do 100 percent inspection.
↑ = Use first sampling plan above arrow.
Ac = Acceptance number
Re = Rejection number
• = Use corresponding single sampling plan (or alternatively, use double sampling plan below, where available).

DOUBLE NORMAL

TABLE III-B — Double sampling plans for tightened inspection (Master table)

(See 9.4 and 9.5)

Sample size code letter	Sample	Sample size	Cumulative sample size	Acceptable Quality Levels (tightened inspection)																									
				0.010	0.015	0.025	0.040	0.065	0.10	0.15	0.25	0.40	0.65	1.0	1.5	2.5	4.0	6.5	10	15	25	40	65	100	150	250	400	650	1000
				Ac Re	Ac Re	Ac Re	Ac Re	Ac Re	Ac Re	Ac Re	Ac Re	Ac Re	Ac Re	Ac Re	Ac Re	Ac Re	Ac Re	Ac Re	Ac Re	Ac Re	Ac Re	Ac Re	Ac Re	Ac Re	Ac Re	Ac Re	Ac Re	Ac Re	Ac Re
A				↓	↓	↓	↓	↓	↓	↓	↓	↓	↓	↓	↓	↓	↓	↓	↓	↓	↓	•	•	•	•	•	•	•	•
B	First	2	2	↓	↓	↓	↓	↓	↓	↓	↓	↓	↓	↓	↓	↓	↓	•	↓	↓	0 2	0 3	1 4	2 5	3 7	6 10	9 14	15 20	23 29
	Second	2	4																		1 2	3 4	4 5	6 7	11 12	15 16	23 24	34 35	52 53
C	First	3	3	↓	↓	↓	↓	↓	↓	↓	↓	↓	↓	↓	↓	↓	•	↓	↓	0 2	0 3	1 4	2 5	3 7	6 10	9 14	15 20	23 29	↑
	Second	3	6																	1 2	3 4	4 5	6 7	11 12	15 16	23 24	34 35	52 53	
D	First	5	5	↓	↓	↓	↓	↓	↓	↓	↓	↓	↓	↓	↓	•	↓	↓	0 2	0 3	1 4	2 5	3 7	6 10	9 14	15 20	23 29	↑	↑
	Second	5	10																1 2	3 4	4 5	6 7	11 12	15 16	23 24	34 35	52 53		
E	First	8	8	↓	↓	↓	↓	↓	↓	↓	↓	↓	↓	↓	•	↓	↓	0 2	0 3	1 4	2 5	3 7	6 10	9 14	15 20	23 29	↑	↑	↑
	Second	8	16															1 2	3 4	4 5	6 7	11 12	15 16	23 24	34 35	52 53			
F	First	13	13	↓	↓	↓	↓	↓	↓	↓	↓	↓	↓	•	↓	↓	0 2	0 3	1 4	2 5	3 7	6 10	9 14	↑	↑	↑	↑	↑	↑
	Second	13	26														1 2	3 4	4 5	6 7	11 12	15 16	23 24						
G	First	20	20	↓	↓	↓	↓	↓	↓	↓	↓	↓	•	↓	↓	0 2	0 3	1 4	2 5	3 7	6 10	9 14	↑	↑	↑	↑	↑	↑	↑
	Second	20	40													1 2	3 4	4 5	6 7	11 12	15 16	23 24							
H	First	32	32	↓	↓	↓	↓	↓	↓	↓	↓	•	↓	↓	0 2	0 3	1 4	2 5	3 7	6 10	9 14	↑	↑	↑	↑	↑	↑	↑	↑
	Second	32	64												1 2	3 4	4 5	6 7	11 12	15 16	23 24								
J	First	50	50	↓	↓	↓	↓	↓	↓	↓	•	↓	↓	0 2	0 3	1 4	2 5	3 7	6 10	9 14	↑	↑	↑	↑	↑	↑	↑	↑	↑
	Second	50	100											1 2	3 4	4 5	6 7	11 12	15 16	23 24									
K	First	80	80	↓	↓	↓	↓	↓	↓	•	↓	↓	0 2	0 3	1 4	2 5	3 7	6 10	9 14	↑	↑	↑	↑	↑	↑	↑	↑	↑	↑
	Second	80	160										1 2	3 4	4 5	6 7	11 12	15 16	23 24										
L	First	125	125	↓	↓	↓	↓	↓	•	↓	↓	0 2	0 3	1 4	2 5	3 7	6 10	9 14	↑	↑	↑	↑	↑	↑	↑	↑	↑	↑	↑
	Second	125	250									1 2	3 4	4 5	6 7	11 12	15 16	23 24											
M	First	200	200	↓	↓	↓	↓	•	↓	↓	0 2	0 3	1 4	2 5	3 7	6 10	9 14	↑	↑	↑	↑	↑	↑	↑	↑	↑	↑	↑	↑
	Second	200	400								1 2	3 4	4 5	6 7	11 12	15 16	23 24												
N	First	315	315	↓	↓	↓	•	↓	↓	0 2	0 3	1 4	2 5	3 7	6 10	9 14	↑	↑	↑	↑	↑	↑	↑	↑	↑	↑	↑	↑	↑
	Second	315	630							1 2	3 4	4 5	6 7	11 12	15 16	23 24													
P	First	500	500	↓	↓	•	↓	↓	0 2	0 3	1 4	2 5	3 7	6 10	9 14	↑	↑	↑	↑	↑	↑	↑	↑	↑	↑	↑	↑	↑	↑
	Second	500	1000						1 2	3 4	4 5	6 7	11 12	15 16	23 24														
Q	First	800	800	↓	•	↓	↓	0 2	0 3	1 4	2 5	3 7	6 10	9 14	↑	↑	↑	↑	↑	↑	↑	↑	↑	↑	↑	↑	↑	↑	↑
	Second	800	1600					1 2	3 4	4 5	6 7	11 12	15 16	23 24															
R	First	1250	1250	•	↑	↓	0 2	0 3	1 4	2 5	3 7	6 10	9 14	↑	↑	↑	↑	↑	↑	↑	↑	↑	↑	↑	↑	↑	↑	↑	↑
	Second	1250	2500				1 2	3 4	4 5	6 7	11 12	15 16	23 24																
S	First	2000	2000			0 2																							
	Second	2000	4000			1 2																							

↓ = Use first sampling plan below arrow. If sample size equals or exceeds lot or batch size, do 100 percent inspection.
↑ = Use first sampling plan above arrow.
Ac = Acceptance number
Re = Rejection number
• = Use corresponding single sampling plan (or, alternatively, use double sampling plan below, where available).

DOUBLE TIGHTENED

TABLE III-C — Double sampling plans for reduced inspection (Master table)

(See 9.4 and 9.5)

Sample size code letter	Sample	Sample size	Cumulative sample size	Acceptable Quality Levels (reduced inspection)†																									
				0.010	0.015	0.025	0.040	0.065	0.10	0.15	0.25	0.40	0.65	1.0	1.5	2.5	4.0	6.5	10	15	25	40	65	100	150	250	400	650	1000
				Ac Re	Ac Re	Ac Re	Ac Re	Ac Re	Ac Re	Ac Re	Ac Re	Ac Re	Ac Re	Ac Re	Ac Re	Ac Re	Ac Re	Ac Re	Ac Re	Ac Re	Ac Re	Ac Re	Ac Re	Ac Re	Ac Re	Ac Re	Ac Re	Ac Re	Ac Re
A				↓	↓	↓	↓	↓	↓	↓	↓	↓	↓	↓	↓	↓	↓	•	↓	↓	•	•	•	•	•	•	•	•	•
B				↓	↓	↓	↓	↓	↓	↓	↓	↓	↓	↓	↓	↓	•	↑	↓	•	•	•	•	•	•	•	•	•	•
C				↓	↓	↓	↓	↓	↓	↓	↓	↓	↓	↓	↓	•	↑	↓	•	•	•	•	•	•	•	•	•	•	↑
D	First	2	2	↓	↓	↓	↓	↓	↓	↓	↓	↓	↓	↓	•	↑	↓	0 2	0 3	0 4	0 4	1 5	2 7	3 8	5 10	7 12	11 17	↑	↑
	Second	2	4															0 2	0 4	1 5	3 6	4 7	6 9	8 12	12 16	18 22	26 30		
E	First	3	3	↓	↓	↓	↓	↓	↓	↓	↓	↓	↓	•	↑	↓	0 2	0 3	0 4	0 4	1 5	2 7	3 8	5 10	7 12	11 17	↑	↑	↑
	Second	3	6														0 2	0 4	1 5	3 6	4 7	6 9	8 12	12 16	18 22	26 30			
F	First	5	5	↓	↓	↓	↓	↓	↓	↓	↓	↓	•	↑	↓	0 2	0 3	0 4	0 4	1 5	2 7	3 8	5 10	↑	↑	↑	↑	↑	↑
	Second	5	10													0 2	0 4	1 5	3 6	4 7	6 9	8 12	12 16						
G	First	8	8	↓	↓	↓	↓	↓	↓	↓	↓	•	↑	↓	0 2	0 3	0 4	0 4	1 5	2 7	3 8	5 10	↑	↑	↑	↑	↑	↑	↑
	Second	8	16												0 2	0 4	1 5	3 6	4 7	6 9	8 12	12 16							
H	First	13	13	↓	↓	↓	↓	↓	↓	↓	•	↑	↓	0 2	0 3	0 4	0 4	1 5	2 7	3 8	5 10	↑	↑	↑	↑	↑	↑	↑	↑
	Second	13	26											0 2	0 4	1 5	3 6	4 7	6 9	8 12	12 16								
J	First	20	20	↓	↓	↓	↓	↓	↓	•	↑	↓	0 2	0 3	0 4	0 4	1 5	2 7	3 8	5 10	↑	↑	↑	↑	↑	↑	↑	↑	↑
	Second	20	40										0 2	0 4	1 5	3 6	4 7	6 9	8 12	12 16									
K	First	32	32	↓	↓	↓	↓	↓	•	↑	↓	0 2	0 3	0 4	0 4	1 5	2 7	3 8	5 10	↑	↑	↑	↑	↑	↑	↑	↑	↑	↑
	Second	32	64									0 2	0 4	1 5	3 6	4 7	6 9	8 12	12 16										
L	First	50	50	↓	↓	↓	↓	•	↑	↓	0 2	0 3	0 4	0 4	1 5	2 7	3 8	5 10	↑	↑	↑	↑	↑	↑	↑	↑	↑	↑	↑
	Second	50	100								0 2	0 4	1 5	3 6	4 7	6 9	8 12	12 16											
M	First	80	80	↓	↓	↓	•	↑	↓	0 2	0 3	0 4	0 4	1 5	2 7	3 8	5 10	↑	↑	↑	↑	↑	↑	↑	↑	↑	↑	↑	↑
	Second	80	160							0 2	0 4	1 5	3 6	4 7	6 9	8 12	12 16												
N	First	125	125	↓	↓	•	↑	↓	0 2	0 3	0 4	0 4	1 5	2 7	3 8	5 10	↑	↑	↑	↑	↑	↑	↑	↑	↑	↑	↑	↑	↑
	Second	125	250						0 2	0 4	1 5	3 6	4 7	6 9	8 12	12 16													
P	First	200	200	↓	•	↑	↓	0 2	0 3	0 4	0 4	1 5	2 7	3 8	5 10	↑	↑	↑	↑	↑	↑	↑	↑	↑	↑	↑	↑	↑	↑
	Second	200	400					0 2	0 4	1 5	3 6	4 7	6 9	8 12	12 16														
Q	First	315	315	•	↑	↓	0 2	0 3	0 4	0 4	1 5	2 7	3 8	5 10	↑	↑	↑	↑	↑	↑	↑	↑	↑	↑	↑	↑	↑	↑	↑
	Second	315	630				0 2	0 4	1 5	3 6	4 7	6 9	8 12	12 16															
R	First	500	500	↑	↑	0 2	0 3	0 4	0 4	1 5	2 7	3 8	5 10	↑	↑	↑	↑	↑	↑	↑	↑	↑	↑	↑	↑	↑	↑	↑	↑
	Second	500	1000			0 2	0 4	1 5	3 6	4 7	6 9	8 12	12 16																

↓ = Use first sampling plan below arrow. If sample size equals or exceeds lot or batch size, do 100 percent inspection.
↑ = Use first sampling plan above arrow.
Ac = Acceptance number.
Re = Rejection number.
• = Use corresponding single sampling plan (or alternatively, use double sampling plan below, when available.)
† = If, after the second sample, the acceptance number has been exceeded, but the rejection number has not been reached, accept the lot, but reinstate normal inspection (see 10.14).

TABLE IV-A — *Multiple sampling plans for normal inspection (Master table)*

(See 9.4 and 9.5)

Sample size code letter	Sample	Sample size	Cumulative sample size	0.010	0.015	0.025	0.040	0.065	0.10	0.15	0.25	0.40	0.65	1.0	1.5	2.5	4.0	6.5	10	15	25	40	65	100	150	250	400	650	1000
				Ac Re	Ac Re	Ac Re	Ac Re	Ac Re	Ac Re	Ac Re	Ac Re	Ac Re	Ac Re	Ac Re	Ac Re	Ac Re	Ac Re	Ac Re	Ac Re	Ac Re	Ac Re	Ac Re	Ac Re	Ac Re	Ac Re	Ac Re	Ac Re	Ac Re	Ac Re
A				↓	↓	↓	↓	↓	↓	↓	↓	↓	↓	↓	↓	↓	↓	•	↓	↓	•	•	•	•	•	•	•	•	•
B				↓	↓	↓	↓	↓	↓	↓	↓	↓	↓	↓	↓	↓	•	↑	↓	‡	‡	‡	‡	‡	‡	‡	‡	‡	‡
C				↓	↓	↓	↓	↓	↓	↓	↓	↓	↓	↓	↓	•	↑	↓	‡	‡	‡	‡	‡	‡	‡	‡	‡	‡	↑
D	First	2	2	↓	↓	↓	↓	↓	↓	↓	↓	↓	↓	↓	•	↑	↓	* 2	* 2	* 3	* 4	0 4	0 5	1 7	2 9	4 12	6 16	↑	↑
	Second	2	4															* 2	0 3	0 3	1 5	1 6	3 8	4 10	7 14	11 19	17 27		
	Third	2	6															0 2	0 3	1 4	2 6	3 8	6 10	8 13	13 19	19 27	29 39		
	Fourth	2	8															0 3	1 4	2 5	3 7	5 10	8 13	12 17	19 25	27 34	40 49		
	Fifth	2	10															1 3	2 4	3 6	5 8	7 11	11 15	17 20	25 29	36 40	53 58		
	Sixth	2	12															1 3	3 5	4 6	7 9	10 12	14 17	21 23	31 33	45 47	65 68		
	Seventh	2	14															2 3	4 5	6 7	9 10	13 14	18 19	25 26	37 38	53 54	77 78		
E	First	3	3	↓	↓	↓	↓	↓	↓	↓	↓	↓	↓	•	↑	↓	* 2	* 2	* 3	* 4	0 4	0 5	1 7	2 9	4 12	6 16	↑	↑	↑
	Second	3	6														* 2	0 3	0 3	1 5	1 6	3 8	4 10	7 14	11 19	17 27			
	Third	3	9														0 2	0 3	1 4	2 6	3 8	6 10	8 13	13 19	19 27	29 39			
	Fourth	3	12														0 3	1 4	2 5	3 7	5 10	8 13	12 17	19 25	27 34	40 49			
	Fifth	3	15														1 3	2 4	3 6	5 8	7 11	11 15	17 20	25 29	36 40	53 58			
	Sixth	3	18														1 3	3 5	4 6	7 9	10 12	14 17	21 23	31 33	45 47	65 68			
	Seventh	3	21														2 3	4 5	6 7	9 10	13 14	18 19	25 26	37 38	53 54	77 78			
F	First	5	5	↓	↓	↓	↓	↓	↓	↓	↓	↓	•	↑	↓	* 2	* 2	* 3	* 4	0 4	0 5	1 7	2 9	↑	↑	↑	↑	↑	↑
	Second	5	10													* 2	0 3	0 3	1 5	1 6	3 8	4 10	7 14						
	Third	5	15													0 2	0 3	1 4	2 6	3 8	6 10	8 13	13 19						
	Fourth	5	20													0 3	1 4	2 5	3 7	5 10	8 13	12 17	19 25						
	Fifth	5	25													1 3	2 4	3 6	5 8	7 11	11 15	17 20	25 29						
	Sixth	5	30													1 3	3 5	4 6	7 9	10 12	14 17	21 23	31 33						
	Seventh	5	35													2 3	4 5	6 7	9 10	13 14	18 19	25 26	37 38						
G	First	8	8	↓	↓	↓	↓	↓	↓	↓	↓	•	↑	↓	* 2	* 2	* 3	* 4	0 4	0 5	1 7	2 9	↑	↑	↑	↑	↑	↑	↑
	Second	8	16												* 2	0 3	0 3	1 5	1 6	3 8	4 10	7 14							
	Third	8	24												0 2	0 3	1 4	2 6	3 8	6 10	8 13	13 19							
	Fourth	8	32												0 3	1 4	2 5	3 7	5 10	8 13	12 17	19 25							
	Fifth	8	40												1 3	2 4	3 6	5 8	7 11	11 15	17 20	25 29							
	Sixth	8	48												1 3	3 5	4 6	7 9	10 12	14 17	21 23	31 33							
	Seventh	8	56												2 3	4 5	6 7	9 10	13 14	18 19	25 26	37 38							
H	First	13	13	↓	↓	↓	↓	↓	↓	↓	•	↑	↓	* 2	* 2	* 3	* 4	0 4	0 5	1 7	2 9	↑	↑	↑	↑	↑	↑	↑	↑
	Second	13	26											* 2	0 3	0 3	1 5	1 6	3 8	4 10	7 14								
	Third	13	39											0 2	0 3	1 4	2 6	3 8	6 10	8 13	13 19								
	Fourth	13	52											0 3	1 4	2 5	3 7	5 10	8 13	12 17	19 25								
	Fifth	13	65											1 3	2 4	3 6	5 8	7 11	11 15	17 20	25 29								
	Sixth	13	78											1 3	3 5	4 6	7 9	10 12	14 17	21 23	31 33								
	Seventh	13	91											2 3	4 5	6 7	9 10	13 14	18 19	25 26	37 38								
J	First	20	20	↓	↓	↓	↓	↓	↓	•	↑	↓	* 2	* 2	* 3	* 4	0 4	0 5	1 7	2 9	↑	↑	↑	↑	↑	↑	↑	↑	↑
	Second	20	40										* 2	0 3	0 3	1 5	1 6	3 8	4 10	7 14									
	Third	20	60										0 2	0 3	1 4	2 6	3 8	6 10	8 13	13 19									
	Fourth	20	80										0 3	1 4	2 5	3 7	5 10	8 13	12 17	19 25									
	Fifth	20	100										1 3	2 4	3 6	5 8	7 11	11 15	17 20	25 29									
	Sixth	20	120										1 3	3 5	4 6	7 9	10 12	14 17	21 23	31 33									
	Seventh	20	140										2 3	4 5	6 7	9 10	13 14	18 19	25 26	37 38									

↓ = Use first sampling plan below arrow (refer to continuation of table on following page, when necessary). If sample size equals or exceeds lot or batch size, do 100 percent inspection.
↑ = Use first sampling plan above arrow.
Ac = Acceptance number.
Re = Rejection number.
• = Use corresponding single sampling plan (or alternatively, use multiple sampling plan below, where available).
‡ = Use corresponding double sampling plan (or alternatively, use multiple sampling plan below, where available).
* = Acceptance not permitted at this sample size.

MULTIPLE
NORMAL

TABLE IV-A — *Multiple sampling plans for normal inspection (Master table)* *(Continued)*

(See 9.4 and 9.5)

Sample size code letter	Sample	Sample size	Cumulative sample size	Acceptable Quality Levels (normal inspection)																									
				0.010	0.015	0.025	0.040	0.065	0.10	0.15	0.25	0.40	0.65	1.0	1.5	2.5	4.0	6.5	10	15	25	40	65	100	150	250	400	650	1000
				Ac Re	Ac Re	Ac Re	Ac Re	Ac Re	Ac Re	Ac Re	Ac Re	Ac Re	Ac Re	Ac Re	Ac Re	Ac Re	Ac Re	Ac Re	Ac Re	Ac Re	Ac Re	Ac Re	Ac Re	Ac Re	Ac Re	Ac Re	Ac Re	Ac Re	Ac Re
K	First	32	32	↓	↓	↓	↓	↓	•	↑	↓	* 2	* 2	* 3	* 4	0 4	0 5	1 7	2 9	↑	↑	↑	↑	↑	↑	↑	↑	↑	↑
	Second	32	64									* 2	0 3	0 3	1 5	1 6	3 8	4 10	7 14										
	Third	32	96									0 2	0 3	1 4	2 6	3 8	6 10	8 13	13 19										
	Fourth	32	128									0 3	1 4	2 5	3 7	5 10	8 13	12 17	19 25										
	Fifth	32	160									1 3	2 4	3 6	5 8	7 11	11 15	17 20	25 29										
	Sixth	32	192									1 3	3 5	4 6	7 9	10 12	14 17	21 23	31 33										
	Seventh	32	224									2 3	4 5	6 7	9 10	13 14	18 19	25 26	37 38										
L	First	50	50	↓	↓	↓	↓	•	↑	↓	* 2	* 2	* 3	* 4	0 4	0 5	1 7	2 9	↑	↑	↑	↑	↑	↑	↑	↑	↑	↑	↑
	Second	50	100								* 2	0 3	0 3	1 5	1 6	3 8	4 10	7 14											
	Third	50	150								0 2	0 3	1 4	2 6	3 8	6 10	8 13	13 19											
	Fourth	50	200								0 3	1 4	2 5	3 7	5 10	8 13	12 17	19 25											
	Fifth	50	250								1 3	2 4	3 6	5 8	7 11	11 15	17 20	25 29											
	Sixth	50	300								1 3	3 5	4 6	7 9	10 12	14 17	21 23	31 33											
	Seventh	50	350								2 3	4 5	6 7	9 10	13 14	18 19	25 26	37 38											
M	First	80	80	↓	↓	↓	•	↑	↓	* 2	* 2	* 3	* 4	0 4	0 5	1 7	2 9	↑	↑	↑	↑	↑	↑	↑	↑	↑	↑	↑	↑
	Second	80	160							* 2	0 3	0 3	1 5	1 6	3 8	4 10	7 14												
	Third	80	240							0 2	0 3	1 4	2 6	3 8	6 10	8 13	13 19												
	Fourth	80	320							0 3	1 4	2 5	3 7	5 10	8 13	12 17	19 25												
	Fifth	80	400							1 3	2 4	3 6	5 8	7 11	11 15	17 20	25 29												
	Sixth	80	480							1 3	3 5	4 6	7 9	10 12	14 17	21 23	31 33												
	Seventh	80	560							2 3	4 5	6 7	9 10	13 14	18 19	25 26	37 38												
N	First	125	125	↓	↓	•	↑	↓	* 2	* 2	* 3	* 4	0 4	0 5	1 7	2 9	↑	↑	↑	↑	↑	↑	↑	↑	↑	↑	↑	↑	↑
	Second	125	250						* 2	0 3	0 3	1 5	1 6	3 8	4 10	7 11													
	Third	125	375						0 2	0 3	1 4	2 6	3 8	6 10	8 13	13 19													
	Fourth	125	500						0 3	1 4	2 5	3 7	5 10	8 13	12 17	19 25													
	Fifth	125	625						1 3	2 4	3 6	5 8	7 11	11 15	17 20	25 29													
	Sixth	125	750						1 3	3 5	4 6	7 9	10 12	14 17	21 23	31 33													
	Seventh	125	875						2 3	4 5	6 7	9 10	13 14	18 19	25 26	37 38													
P	First	200	200	↓	•	↑	↓	* 2	* 2	* 3	* 4	0 4	0 5	1 7	2 9	↑	↑	↑	↑	↑	↑	↑	↑	↑	↑	↑	↑	↑	↑
	Second	200	400					* 2	0 3	0 3	1 5	1 6	3 8	4 10	7 14														
	Third	200	600					0 2	0 3	1 4	2 6	3 8	6 10	8 13	13 19														
	Fourth	200	800					0 3	1 4	2 5	3 7	5 10	8 13	12 17	19 25														
	Fifth	200	1000					1 3	2 4	3 6	5 8	7 11	11 15	17 20	25 29														
	Sixth	200	1200					1 3	3 5	4 6	7 9	10 12	14 17	21 23	31 33														
	Seventh	200	1400					2 3	4 5	6 7	9 10	13 14	18 19	25 26	37 38														
Q	First	315	315	•	↑	↓	* 2	* 2	* 3	* 4	0 4	0 5	1 7	2 9	↑	↑	↑	↑	↑	↑	↑	↑	↑	↑	↑	↑	↑	↑	↑
	Second	315	630				* 2	0 3	0 3	1 5	1 6	3 8	4 10	7 14															
	Third	315	945				0 2	0 3	1 4	2 6	3 8	6 10	8 13	13 19															
	Fourth	315	1260				0 3	1 4	2 5	3 7	5 10	8 13	12 17	19 25															
	Fifth	315	1575				1 3	2 4	3 6	5 8	7 11	11 15	17 20	25 29															
	Sixth	315	1890				1 3	3 5	4 6	7 9	10 12	14 17	21 23	31 33															
	Seventh	315	2205				2 3	4 5	6 7	9 10	13 14	18 19	25 26	37 38															
R	First	500	500	↑	↑	* 2	* 2	* 3	* 4	0 4	0 5	1 7	2 9	↑	↑	↑	↑	↑	↑	↑	↑	↑	↑	↑	↑	↑	↑	↑	↑
	Second	500	1000			* 2	0 3	0 3	1 5	1 6	3 8	4 10	7 14																
	Third	500	1500			0 2	0 3	1 4	2 6	3 8	6 10	8 13	13 19																
	Fourth	500	2000			0 3	1 4	2 5	3 7	5 10	8 13	12 17	19 25																
	Fifth	500	2500			1 3	2 4	3 6	5 8	7 11	11 15	17 20	25 29																
	Sixth	500	3000			1 3	3 5	4 6	7 9	10 12	14 17	21 23	31 33																
	Seventh	500	3500			2 3	4 5	6 7	9 10	13 14	18 19	25 26	37 38																

↓ = Use first sampling plan below arrow. If sample size equals or exceeds lot or batch size, do 100 percent inspection.
↑ = Use first sampling plan above arrow (refer to preceding page, when necessary).
Ac = Acceptance number.
Re = Rejection number.
• = Use corresponding single sampling plan (or alternatively, use multiple plan below, where available).
* = Acceptance not permitted at this sample size.

MULTIPLE
NORMAL

MULTIPLE
TIGHTENED

TABLE IV-B—Multiple sampling plans for tightened inspection (Master table)

(See 9.4 and 9.5)

Sample size code letter	Sample	Sample size	Cumulative sample size	0.010	0.015	0.025	0.040	0.065	0.10	0.15	0.25	0.40	0.65	1.0	1.5	2.5	4.0	6.5	10	15	25	40	65	100	150	250	400	650	1000
				Ac Re	Ac Re	Ac Re	Ac Re	Ac Re	Ac Re	Ac Re	Ac Re	Ac Re	Ac Re	Ac Re	Ac Re	Ac Re	Ac Re	Ac Re	Ac Re	Ac Re	Ac Re	Ac Re	Ac Re	Ac Re	Ac Re	Ac Re	Ac Re	Ac Re	Ac Re
A				↓	↓	↓	↓	↓	↓	↓	↓	↓	↓	↓	↓	↓	↓	⇩	↓	↓	⇩	•	•	•	•	•	•	•	•
B				↓	↓	↓	↓	↓	↓	↓	↓	↓	↓	↓	↓	↓	↓	•	↓	↓	↔	↔	↔	↔	↔	↔	↔	↔	↔
C				↓	↓	↓	↓	↓	↓	↓	↓	↓	↓	↓	↓	↓	•	↓	↓	↔	↔	↔	↔	↔	↔	↔	↔	↔	↑
D	First	2	2	↓	↓	↓	↓	↓	↓	↓	↓	↓	↓	↓	↓	•	↓	↓	* 2	* 2	* 3	* 4	0 4	0 6	1 8	3 10	6 15	↑	↑
	Second	2	4																* 2	0 3	0 3	1 5	2 7	3 9	6 12	10 17	16 25		
	Third	2	6																0 2	0 3	1 4	2 6	4 9	7 12	11 17	17 24	26 36		
	Fourth	2	8																0 3	1 4	2 5	3 7	6 11	10 15	16 22	24 31	37 46		
	Fifth	2	10																1 3	2 4	3 6	5 8	9 12	14 17	22 25	32 37	49 55		
	Sixth	2	12																1 3	3 5	4 6	7 9	12 14	18 20	27 29	40 43	61 64		
	Seventh	2	14																2 3	4 5	6 7	9 10	14 15	21 22	32 33	48 49	72 73		
E	First	3	3	↓	↓	↓	↓	↓	↓	↓	↓	↓	↓	↓	•	↓	↓	* 2	* 2	* 3	* 4	0 4	0 6	1 8	3 10	6 15	↑	↑	↑
	Second	3	6															* 2	0 3	0 3	1 5	2 7	3 9	6 12	10 17	16 25			
	Third	3	9															0 2	0 3	1 4	2 6	4 9	7 12	11 17	17 24	26 36			
	Fourth	3	12															0 3	1 4	2 5	3 7	6 11	10 15	16 22	24 31	37 46			
	Fifth	3	15															1 3	2 4	3 6	5 8	9 12	14 17	22 25	32 37	49 55			
	Sixth	3	18															1 3	3 5	4 6	7 9	12 14	18 20	27 29	40 43	61 64			
	Seventh	3	21															2 3	4 5	6 7	9 10	14 15	21 22	32 33	48 49	72 73			
F	First	5	5	↓	↓	↓	↓	↓	↓	↓	↓	↓	↓	•	↓	↓	* 2	* 2	* 3	* 4	0 4	0 6	1 8	↑	↑	↑	↑	↑	↑
	Second	5	10														* 2	0 3	0 3	1 5	2 7	3 9	6 12						
	Third	5	15														0 2	0 3	1 4	2 6	4 9	7 12	11 17						
	Fourth	5	20														0 3	1 4	2 5	3 7	6 11	10 15	16 22						
	Fifth	5	25														1 3	2 4	3 6	5 8	9 12	14 17	22 25						
	Sixth	5	30														1 3	3 5	4 6	7 9	12 14	18 20	27 29						
	Seventh	5	35														2 3	4 5	6 7	9 10	14 15	21 22	32 33						
G	First	8	8	↓	↓	↓	↓	↓	↓	↓	↓	↓	•	↓	↓	* 2	* 2	* 3	* 4	0 4	0 6	1 8	↑	↑	↑	↑	↑	↑	↑
	Second	8	16													* 2	0 3	0 3	1 5	2 7	3 9	6 12							
	Third	8	24													0 2	0 3	1 4	2 6	4 9	7 12	11 17							
	Fourth	8	32													0 3	1 4	2 5	3 7	6 11	10 15	16 22							
	Fifth	8	40													1 3	2 4	3 6	5 8	9 12	14 17	22 25							
	Sixth	8	48													1 3	3 5	4 6	7 9	12 14	18 20	27 29							
	Seventh	8	56													2 3	4 5	6 7	9 10	14 15	21 22	32 33							
H	First	13	13	↓	↓	↓	↓	↓	↓	↓	↓	•	↓	↓	* 2	* 2	* 3	* 4	0 4	0 6	1 8	↑	↑	↑	↑	↑	↑	↑	↑
	Second	13	26											* 2	0 3	0 3	1 5	2 7	3 9	6 12									
	Third	13	39											0 2	0 3	1 4	2 6	4 9	7 12	11 17									
	Fourth	13	52											0 3	1 4	2 5	3 7	6 11	10 15	16 22									
	Fifth	13	65											1 3	2 4	3 6	5 8	9 12	14 17	22 25									
	Sixth	13	78											1 3	3 5	4 6	7 9	12 14	18 20	27 29									
	Seventh	13	91											2 3	4 5	6 7	9 10	14 15	21 22	32 33									
J	First	20	20	↓	↓	↓	↓	↓	↓	↓	•	↓	↓	* 2	* 2	* 3	* 4	0 4	0 6	1 8	↑	↑	↑	↑	↑	↑	↑	↑	↑
	Second	20	40										* 2	0 3	0 3	1 5	2 7	3 9	6 12										
	Third	20	60										0 2	0 3	1 4	2 6	4 9	7 12	11 17										
	Fourth	20	80										0 3	1 4	2 5	3 7	6 11	10 15	16 22										
	Fifth	20	100										1 3	2 4	3 6	5 8	9 12	14 17	22 25										
	Sixth	20	120										1 3	3 5	4 6	7 9	12 14	18 20	27 29										
	Seventh	20	140										2 3	4 5	6 7	9 10	14 15	21 22	32 33										

↓ = Use first sampling plan below arrow (refer to continuation of table on following page, when necessary). If sample size equals or exceeds lot or batch size, do 100 percent inspection.
↑ = Use first sampling plan above arrow.
Ac = Acceptance number
Re = Rejection number
• = Use corresponding single sampling plan (or alternatively, use multiple sampling plan below, where available).
↔ = Use corresponding double sampling plan (or alternatively, use multiple sampling plan below, where available).
* = Acceptance not permitted at this sample size.

TABLE IV-B — Multiple sampling plans for tightened inspection (Master table) (Continued)

(See 9.4 and 9.5)

Sample size code letter	Sample	Sample size	Cumulative sample size	Acceptable Quality Levels (tightened inspection)																									
				0.010	0.015	0.025	0.040	0.065	0.10	0.15	0.25	0.40	0.65	1.0	1.5	2.5	4.0	6.5	10	15	25	40	65	100	150	250	400	650	1000
				Ac Re	Ac Re	Ac Re	Ac Re	Ac Re	Ac Re	Ac Re	Ac Re	Ac Re	Ac Re	Ac Re	Ac Re	Ac Re	Ac Re	Ac Re	Ac Re	Ac Re	Ac Re	Ac Re	Ac Re	Ac Re	Ac Re	Ac Re	Ac Re	Ac Re	Ac Re
K	First	32	32	↓	↓	↓	↓	↓	↓	•	↓	↓	* 2	* 2	* 3	* 4	0 4	0 6	1 8	↑	↑	↑	↑	↑	↑	↑	↑	↑	↑
	Second	32	64										* 2	0 3	0 3	1 5	2 7	3 9	6 12										
	Third	32	96										0 2	0 3	1 4	2 6	4 9	7 12	11 17										
	Fourth	32	128										0 3	1 4	2 5	3 7	6 11	10 15	16 22										
	Fifth	32	160										1 3	2 4	3 6	5 8	9 12	14 17	22 25										
	Sixth	32	192										1 3	3 5	4 6	7 9	12 14	18 20	27 29										
	Seventh	32	224										2 3	4 5	6 7	9 10	14 15	21 22	32 33										
L	First	50	50	↓	↓	↓	↓	↓	•	↓	↓	* 2	* 2	* 3	* 4	0 4	0 6	1 8	↑	↑	↑	↑	↑	↑	↑	↑	↑	↑	↑
	Second	50	100									* 2	0 3	0 3	1 5	2 7	3 9	6 12											
	Third	50	150									0 2	0 3	1 4	2 6	4 9	7 12	11 17											
	Fourth	50	200									0 3	1 4	2 5	3 7	6 11	10 15	16 22											
	Fifth	50	250									1 3	2 4	3 6	5 8	9 12	14 17	22 25											
	Sixth	50	300									1 3	3 5	4 6	7 9	12 14	18 20	27 29											
	Seventh	50	350									2 3	4 5	6 7	9 10	14 15	21 22	32 33											
M	First	80	80	↓	↓	↓	↓	•	↓	↓	* 2	* 2	* 3	* 4	0 4	0 6	1 8	↑	↑	↑	↑	↑	↑	↑	↑	↑	↑	↑	↑
	Second	80	160								* 2	0 3	0 3	1 5	2 7	3 9	6 12												
	Third	80	240								0 2	0 3	1 4	2 6	4 9	7 12	11 17												
	Fourth	80	320								0 3	1 4	2 5	3 7	6 11	10 15	16 22												
	Fifth	80	400								-1 3	2 4	3 6	5 8	9 12	14 17	22 25												
	Sixth	80	480								1 3	3 5	4 6	7 9	12 14	18 20	27 29												
	Seventh	80	560								2 3	4 5	6 7	9 10	14 15	21 22	32 33												
N	First	125	125	↓	↓	↓	•	↓	↓	* 2	* 2	* 3	* 4	0 4	0 6	1 8	↑	↑	↑	↑	↑	↑	↑	↑	↑	↑	↑	↑	↑
	Second	125	250							* 2	0 3	0 3	1 5	2 7	3 9	6 12													
	Third	125	375							0 2	0 3	1 4	2 6	4 9	7 12	11 17													
	Fourth	125	500							0 3	1 4	2 5	3 7	6 11	10 15	16 22													
	Fifth	125	625							1 3	2 4	3 6	5 8	9 12	14 17	22 25													
	Sixth	125	750							1 3	3 5	4 6	7 9	12 14	18 20	27 29													
	Seventh	125	875							2 3	4 5	6 7	9 10	14 15	21 22	32 33													
P	First	200	200	↓	↓	•	↓	↓	* 2	* 2	* 3	* 4	0 4	0 6	1 8	↑	↑	↑	↑	↑	↑	↑	↑	↑	↑	↑	↑	↑	↑
	Second	200	400						* 2	0 3	0 3	1 5	2 7	3 9	6 12														
	Third	200	600						0 2	0 3	1 4	2 6	4 9	7 12	11 17														
	Fourth	200	800						0 3	1 4	2 5	3 7	6 11	10 15	16 22														
	Fifth	200	1000						1 3	2 4	3 6	5 8	9 12	14 17	22 25														
	Sixth	200	1200						1 3	3 5	4 6	7 9	12 14	18 20	27 29														
	Seventh	200	1400						2 3	4 5	6 7	9 10	14 15	21 22	32 33														
Q	First	315	315	↓	•	↓	↓	* 2	* 2	* 3	* 4	0 4	0 6	1 8	↑	↑	↑	↑	↑	↑	↑	↑	↑	↑	↑	↑	↑	↑	↑
	Second	315	630					* 2	0 3	0 3	1 5	2 7	3 9	6 12															
	Third	315	945					0 2	0 3	1 4	2 6	4 9	7 12	11 17															
	Fourth	315	1260					0 3	1 4	2 5	3 7	6 11	10 15	16 22															
	Fifth	315	1575					1 3	2 4	3 6	5 8	9 12	14 17	22 25															
	Sixth	315	1890					1 3	3 5	4 6	7 9	12 14	18 20	27 29															
	Seventh	315	2205					2 3	4 5	6 7	9 10	14 15	21 22	32 33															
R	First	500	500	•	↑	↓	* 2	* 2	* 3	* 4	0 4	0 6	1 8	↑	↑	↑	↑	↑	↑	↑	↑	↑	↑	↑	↑	↑	↑	↑	↑
	Second	500	1000				* 2	0 3	0 3	1 5	2 7	3 9	6 12																
	Third	500	1500				0 2	0 3	1 4	2 6	4 9	7 12	11 17																
	Fourth	500	2000				0 3	1 4	2 5	3 7	6 11	10 15	16 22																
	Fifth	500	2500				1 3	2 4	3 6	5 8	9 12	14 17	22 25																
	Sixth	500	3000				1 3	3 5	4 6	7 9	12 14	18 20	27 29																
	Seventh	500	3500				2 3	4 5	6 7	9 10	14 15	21 22	32 33																
S	First	800	800			* 2																							
	Second	800	1600			* 2																							
	Third	800	2400			0 2																							
	Fourth	800	3200			0 3																							
	Fifth	800	4000			1 3																							
	Sixth	800	4800			1 3																							
	Seventh	800	5600			2 3																							

↓ = Use first sampling plan below arrow. If sample size equals or exceeds lot or batch size, do 100 percent inspection.
↑ = Use first sampling plan above arrow (refer to preceding page, when necessary).
Ac = Acceptance number
Re = Rejection number
• = Use corresponding single sampling plan (or alternatively, use multiple sampling plan below, where available).
* = Acceptance not permitted at this sample size.

MULTIPLE TIGHTENED

TABLE IV-C—*Multiple sampling plans for reduced inspection (Master table)*

(See 9.4 and 9.5)

Sample size code letter	Sample	Sample size	Cumu-lative sample size	Acceptable Quality Levels (reduced inspection) †																									
				0.010	0.015	0.025	0.040	0.065	0.10	0.15	0.25	0.40	0.65	1.0	1.5	2.5	4.0	6.5	10	15	25	40	65	100	150	250	400	650	1000
				Ac Re	Ac Re	Ac Re	Ac Re	Ac Re	Ac Re	Ac Re	Ac Re	Ac Re	Ac Re	Ac Re	Ac Re	Ac Re	Ac Re	Ac Re	Ac Re	Ac Re	Ac Re	Ac Re	Ac Re	Ac Re	Ac Re	Ac Re	Ac Re	Ac Re	Ac Re
A				↓	↓	↓	↓	↓	↓	↓	↓	↓	↓	↓	↓	↓	↓	•	↓	↓	•	•	•	•	•	•	•	•	•
B				↓	↓	↓	↓	↓	↓	↓	↓	↓	↓	↓	↓	↓	•	↑	↓	•	•	•	•	•	•	•	•	•	•
C				↓	↓	↓	↓	↓	↓	↓	↓	↓	↓	↓	↓	•	↑	↓	•	•	•	•	•	•	•	•	•	•	↑
D				↓	↓	↓	↓	↓	↓	↓	↓	↓	↓	↓	•	↑	↓	↔	↔	↔	↔	↔	↔	↔	↔	↔	↔	↑	↑
E				↓	↓	↓	↓	↓	↓	↓	↓	↓	↓	•	↑	↓	↔	↔	↔	↔	↔	↔	↔	↔	↔	↔	↑	↑	↑
F	First	2	2	↓	↓	↓	↓	↓	↓	↓	↓	↓	•	↑	↓	# 2	# 2	# 3	# 3	# 4	# 4	0 5	0 6	↑	↑	↑	↑	↑	↑
	Second	2	4	↓	↓	↓	↓	↓	↓	↓	↓	↓		↑	↓	# 2	# 3	# 3	0 4	0 5	1 6	1 7	3 9	↑	↑	↑	↑	↑	↑
	Third	2	6	↓	↓	↓	↓	↓	↓	↓	↓	↓		↑	↓	0 2	0 3	0 4	0 5	1 6	2 8	3 9	6 12	↑	↑	↑	↑	↑	↑
	Fourth	2	8	↓	↓	↓	↓	↓	↓	↓	↓	↓		↑	↓	0 3	0 4	0 5	1 6	2 7	3 10	5 12	8 15	↑	↑	↑	↑	↑	↑
	Fifth	2	10	↓	↓	↓	↓	↓	↓	↓	↓	↓		↑	↓	0 3	0 4	1 6	2 7	3 8	5 11	7 13	11 17	↑	↑	↑	↑	↑	↑
	Sixth	2	12	↓	↓	↓	↓	↓	↓	↓	↓	↓		↑	↓	0 3	1 5	1 6	3 7	4 9	7 12	10 15	14 20	↑	↑	↑	↑	↑	↑
	Seventh	2	14	↓	↓	↓	↓	↓	↓	↓	↓	↓		↑	↓	1 3	1 5	2 7	4 8	6 10	9 14	13 17	18 22	↑	↑	↑	↑	↑	↑
G	First	3	3	↓	↓	↓	↓	↓	↓	↓	↓	•	↑	↓	# 2	# 2	# 3	# 3	# 4	# 4	0 5	0 6	↑	↑	↑	↑	↑	↑	↑
	Second	3	6	↓	↓	↓	↓	↓	↓	↓	↓		↑	↓	# 2	# 3	# 3	0 4	0 5	1 6	1 7	3 9	↑	↑	↑	↑	↑	↑	↑
	Third	3	9	↓	↓	↓	↓	↓	↓	↓	↓		↑	↓	0 2	0 3	0 4	0 5	1 6	2 8	3 9	6 12	↑	↑	↑	↑	↑	↑	↑
	Fourth	3	12	↓	↓	↓	↓	↓	↓	↓	↓		↑	↓	0 3	0 4	0 5	1 6	2 7	3 10	5 12	8 15	↑	↑	↑	↑	↑	↑	↑
	Fifth	3	15	↓	↓	↓	↓	↓	↓	↓	↓		↑	↓	0 3	0 4	1 6	2 7	3 8	5 11	7 13	11 17	↑	↑	↑	↑	↑	↑	↑
	Sixth	3	18	↓	↓	↓	↓	↓	↓	↓	↓		↑	↓	0 3	1 5	1 6	3 7	4 9	7 12	10 15	14 20	↑	↑	↑	↑	↑	↑	↑
	Seventh	3	21	↓	↓	↓	↓	↓	↓	↓	↓		↑	↓	1 3	1 5	2 7	4 8	6 10	9 14	13 17	18 22	↑	↑	↑	↑	↑	↑	↑
H	First	5	5	↓	↓	↓	↓	↓	↓	↓	•	↑	↓	# 2	# 2	# 3	# 3	# 4	# 4	0 5	0 6	↑	↑	↑	↑	↑	↑	↑	↑
	Second	5	10	↓	↓	↓	↓	↓	↓	↓		↑	↓	# 2	# 3	# 3	0 4	0 5	1 6	1 7	3 9	↑	↑	↑	↑	↑	↑	↑	↑
	Third	5	15	↓	↓	↓	↓	↓	↓	↓		↑	↓	0 2	0 3	0 4	0 5	1 6	2 8	3 9	6 12	↑	↑	↑	↑	↑	↑	↑	↑
	Fourth	5	20	↓	↓	↓	↓	↓	↓	↓		↑	↓	0 3	0 4	0 5	1 6	2 7	3 10	5 12	8 15	↑	↑	↑	↑	↑	↑	↑	↑
	Fifth	5	25	↓	↓	↓	↓	↓	↓	↓		↑	↓	0 3	0 4	1 6	2 7	3 8	5 11	7 13	11 17	↑	↑	↑	↑	↑	↑	↑	↑
	Sixth	5	30	↓	↓	↓	↓	↓	↓	↓		↑	↓	0 3	1 5	1 6	3 7	4 9	7 12	10 15	14 20	↑	↑	↑	↑	↑	↑	↑	↑
	Seventh	5	35	↓	↓	↓	↓	↓	↓	↓		↑	↓	1 3	1 5	2 7	4 8	6 10	9 14	13 17	18 22	↑	↑	↑	↑	↑	↑	↑	↑
J	First	8	8	↓	↓	↓	↓	↓	↓	•	↑	↓	# 2	# 2	# 3	# 3	# 4	# 4	0 5	0 6	↑	↑	↑	↑	↑	↑	↑	↑	↑
	Second	8	16	↓	↓	↓	↓	↓	↓		↑	↓	# 2	# 3	# 3	0 4	0 5	1 6	1 7	3 9	↑	↑	↑	↑	↑	↑	↑	↑	↑
	Third	8	24	↓	↓	↓	↓	↓	↓		↑	↓	0 2	0 3	0 4	0 5	1 6	2 8	3 9	6 12	↑	↑	↑	↑	↑	↑	↑	↑	↑
	Fourth	8	32	↓	↓	↓	↓	↓	↓		↑	↓	0 3	0 4	0 5	1 6	2 7	3 10	5 12	8 15	↑	↑	↑	↑	↑	↑	↑	↑	↑
	Fifth	8	40	↓	↓	↓	↓	↓	↓		↑	↓	0 3	0 4	1 6	2 7	3 8	5 11	7 13	11 17	↑	↑	↑	↑	↑	↑	↑	↑	↑
	Sixth	8	48	↓	↓	↓	↓	↓	↓		↑	↓	0 3	1 5	1 6	3 7	4 9	7 12	10 15	14 20	↑	↑	↑	↑	↑	↑	↑	↑	↑
	Seventh	8	56	↓	↓	↓	↓	↓	↓		↑	↓	1 3	1 5	2 7	4 8	6 10	9 14	13 17	18 22	↑	↑	↑	↑	↑	↑	↑	↑	↑
K	First	13	13	↓	↓	↓	↓	↓	•	↑	↓	# 2	# 2	# 3	# 3	# 4	# 4	0 5	0 6	↑	↑	↑	↑	↑	↑	↑	↑	↑	↑
	Second	13	26	↓	↓	↓	↓	↓		↑	↓	# 2	# 3	# 3	0 4	0 5	1 6	1 7	3 9	↑	↑	↑	↑	↑	↑	↑	↑	↑	↑
	Third	13	39	↓	↓	↓	↓	↓		↑	↓	0 2	0 3	0 4	0 5	1 6	2 8	3 9	6 12	↑	↑	↑	↑	↑	↑	↑	↑	↑	↑
	Fourth	13	52	↓	↓	↓	↓	↓		↑	↓	0 3	0 4	0 5	1 6	2 7	3 10	5 12	8 15	↑	↑	↑	↑	↑	↑	↑	↑	↑	↑
	Fifth	13	65	↓	↓	↓	↓	↓		↑	↓	0 3	0 4	1 6	2 7	3 8	5 11	7 13	11 17	↑	↑	↑	↑	↑	↑	↑	↑	↑	↑
	Sixth	13	78	↓	↓	↓	↓	↓		↑	↓	0 3	1 5	1 6	3 7	4 9	7 12	10 15	14 20	↑	↑	↑	↑	↑	↑	↑	↑	↑	↑
	Seventh	13	91	↓	↓	↓	↓	↓		↑	↓	1 3	1 5	2 7	4 8	6 10	9 14	13 17	18 22	↑	↑	↑	↑	↑	↑	↑	↑	↑	↑

↓ = Use first sampling plan below arrow (refer to continuation of table on following page, when necessary). If sample size equals, or exceeds lot or batch size, do 100 percent inspection.

↑ = Use first sampling plan above arrow.

Ac = Acceptance number

Re = Rejection number

• = Use corresponding single sampling plan (or alternatively, use multiple sampling plan below, where available).

↔ = Use corresponding double sampling plan (or alternatively, use multiple sampling plan below, where available).

= Acceptance not permitted at this sample size.

† = If, after the final sample, the acceptance number has been exceeded, but the rejection number has not been reached, accept the lot but reinstate normal inspection (see 10.1.4).

TABLE IV-C—Multiple sampling plans for reduced inspection (Master table) (Continued)

(See 9.4 and 9.5)

Sample size code letter	Sample	Sample size	Cumulative sample size	Acceptable Quality Levels (reduced inspection)†																									
				0.010	0.015	0.025	0.040	0.065	0.10	0.15	0.25	0.40	0.65	1.0	1.5	2.5	4.0	6.5	10	15	25	40	65	100	150	250	400	650	1000
				Ac Re	Ac Re	Ac Re	Ac Re	Ac Re	Ac Re	Ac Re	Ac Re	Ac Re	Ac Re	Ac Re	Ac Re	Ac Re	Ac Re	Ac Re	Ac Re	Ac Re	Ac Re	Ac Re	Ac Re	Ac Re	Ac Re	Ac Re	Ac Re	Ac Re	Ac Re
L	First	20	20	↓	↓	↓	↓	•	↑	↓	* 2	* 2	* 3	* 3	* 4	* 4	0 5	0 6	↑	↑	↑	↑	↑	↑	↑	↑	↑	↑	↑
	Second	20	40								* 2	* 3	* 3	0 4	0 5	1 6	1 7	3 9											
	Third	20	60								0 2	0 3	0 4	0 5	1 6	2 8	3 9	6 12											
	Fourth	20	80								0 3	0 4	0 5	1 6	2 7	3 10	5 12	8 15											
	Fifth	20	100								0 3	0 4	1 6	2 7	3 8	5 11	7 13	11 17											
	Sixth	20	120								0 3	1 5	1 6	3 7	4 9	7 12	10 15	14 20											
	Seventh	20	140								1 3	1 5	2 7	4 8	6 10	9 14	13 17	18 22											
M	First	32	32				•	↑	↓	* 2	* 2	* 3	* 3	* 4	* 4	0 5	0 6	↑											
	Second	32	64							* 2	* 3	* 3	0 4	0 5	1 6	1 7	3 9												
	Third	32	96							0 2	0 3	0 4	0 5	1 6	2 8	3 9	6 12												
	Fourth	32	128							0 3	0 4	0 5	1 6	2 7	3 10	5 12	8 15												
	Fifth	32	160							0 3	0 4	1 6	2 7	3 8	5 11	7 13	11 17												
	Sixth	32	192							0 3	1 5	1 6	3 7	4 9	7 12	10 15	14 20												
	Seventh	32	224							1 3	1 5	2 7	4 8	6 10	9 14	13 17	18 22												
N	First	50	50			•	↑	↓	* 2	* 2	* 3	* 3	* 4	* 4	0 5	0 6	↑												
	Second	50	100						* 2	* 3	* 3	0 4	0 5	1 6	1 7	3 9													
	Third	50	150						0 2	0 3	0 4	0 5	1 6	2 8	3 9	6 12													
	Fourth	50	200						0 3	0 4	0 5	1 6	2 7	3 10	5 12	8 15													
	Fifth	50	250						0 3	0 4	1 6	2 7	3 8	5 11	7 13	11 17													
	Sixth	50	300						0 3	1 5	1 6	3 7	4 9	7 12	10 15	14 20													
	Seventh	50	350						1 3	1 5	2 7	4 8	6 10	9 14	13 17	18 22													
P	First	80	80		•	↑	↓	* 2	* 2	* 3	* 3	* 4	* 4	0 5	0 6	↑													
	Second	80	160					* 2	* 3	* 3	0 4	0 5	1 6	1 7	3 9														
	Third	80	240					0 2	0 3	0 4	0 5	1 6	2 8	3 9	6 12														
	Fourth	80	320					0 3	0 4	0 5	1 6	2 7	3 10	5 12	8 15														
	Fifth	80	400					0 3	0 4	1 6	2 7	3 8	5 11	7 13	11 17														
	Sixth	80	480					0 3	1 5	1 6	3 7	4 9	7 12	10 15	14 20														
	Seventh	80	560					1 3	1 5	2 7	4 8	6 10	9 14	13 17	18 22														
Q	First	125	125	•	↑	↓	* 2	* 2	* 3	* 3	* 4	* 4	0 5	0 6	↑														
	Second	125	250				* 2	* 3	* 3	0 4	0 5	1 6	1 7	3 9															
	Third	125	375				0 2	0 3	0 4	0 5	1 6	2 8	3 9	6 12															
	Fourth	125	500				0 3	0 4	0 5	1 6	2 7	3 10	5 12	8 15															
	Fifth	125	625				0 3	0 4	1 6	2 7	3 8	5 11	7 13	11 17															
	Sixth	125	750				0 3	1 5	1 6	3 7	4 9	7 12	10 15	14 20															
	Seventh	125	875				1 3	1 5	2 7	4 8	6 10	9 14	13 17	18 22															
R	First	200	200	↑		* 2	* 2	* 3	* 3	* 4	* 4	0 5	0 6	↑															
	Second	200	400			* 2	* 3	* 3	0 4	0 5	1 6	1 7	3 9																
	Third	200	600			0 2	0 3	0 4	0 5	1 6	2 8	3 9	6 12																
	Fourth	200	800			0 3	0 4	0 5	1 6	2 7	3 10	5 12	8 15																
	Fifth	200	1000			0 3	0 4	1 6	2 7	3 8	5 11	7 13	11 17																
	Sixth	200	1200			0 3	1 5	1 6	3 7	4 9	7 12	10 15	14 20																
	Seventh	200	1400			1 3	1 5	2 7	4 8	6 10	9 14	13 17	18 22																

↓ = Use first sampling plan below arrow. If sample size equals, or exceeds, lot or batch size, do 100 percent inspection.
↑ = Use first sampling plan above arrow (refer to preceding page when necessary).
Ac = Acceptance number
Re = Rejection number
* = Acceptance not permitted at this sample size.
† = If, after the final sample, the acceptance number has been exceeded, but the rejection number has not been reached, accept the lot, but reinstate normal inspection (see 10.1.4).

MULTIPLE REDUCED

TABLE V-A—*Average Outgoing Quality Limit Factors for Normal Inspection (Single sampling)*

(See 11.4)

Code Letter	Sample Size	Acceptable Quality Level																									
		0.010	0.015	0.025	0.040	0.065	0.10	0.15	0.25	0.40	0.65	1.0	1.5	2.5	4.0	6.5	10	15	25	40	65	100	150	250	400	650	1000
A	2															18			42	69	97	160	220	330	470	730	1100
B	3														12			28	46	65	110	150	220	310	490	720	1100
C	5													7.4			17	27	39	63	90	130	190	290	430	660	
D	8												4.6			11	17	24	40	56	82	120	180	270	410		
E	13											2.8			6.5	11	15	24	34	50	72	110	170	250			
F	20										1.8			4.2	6.9	9.7	16	22	33	47	73						
G	32									1.2			2.6	4.3	6.1	9.9	14	21	29	46							
H	50								0.74			1.7	2.7	3.9	6.3	9.0	13	19	29								
J	80							0.46			1.1	1.7	2.4	4.0	5.6	8.2	12	18									
K	125						0.29			0.67	1.1	1.6	2.5	3.6	5.2	7.5	12										
L	200					0.18			0.42	0.69	0.97	1.6	2.2	3.3	4.7	7.3											
M	315				0.12			0.27	0.44	0.62	1.00	1.4	2.1	3.0	4.7												
N	500			0.074			0.17	0.27	0.39	0.63	0.90	1.3	1.9	2.9													
P	800		0.046			0.11	0.17	0.24	0.40	0.56	0.82	1.2	1.8														
Q	1250	0.029			0.067	0.11	0.16	0.25	0.36	0.52	0.75	1.2															
R	2000			0.042	0.069	0.097	0.16	0.22	0.33	0.47	0.73																

Note: For the exact AOQL, the above values must be multiplied by $\left(1 - \frac{\text{Sample size}}{\text{Lot or Batch size}}\right)$ (see 11.4)

TABLE V-B—Average Outgoing Quality Limit Factors for Tightened Inspection (Single sampling)

(See 11.4)

Code letter	Sample size	Acceptable Quality Level																									
		0.010	0.015	0.025	0.040	0.065	0.10	0.15	0.25	0.40	0.65	1.0	1.5	2.5	4.0	6.5	10	15	25	40	65	100	150	250	400	650	1000
A	2																			42	69	97	160	260	400	620	970
B	3															12			28	46	65	110	170	270	410	650	1100
C	5														7.4			17	27	39	63	100	160	250	390	610	
D	8													4.6			11	17	24	40	64	99	160	240	380		
E	13												2.8			6.5	11	15	24	40	61	95	150	240			
F	20											1.8			4.2	6.9	9.7	16	26	40	62						
G	32										1.2			2.6	4.3	6.1	9.9	16	25	39							
H	50									0.74			1.7	2.7	3.9	6.3	10	16	25								
J	80								0.46			1.1	1.7	2.4	4.0	6.4	9.9	16									
K	125							0.29			0.67	1.1	1.6	2.5	4.1	6.4	9.9										
L	200						0.18			0.42	0.69	0.97	1.6	2.6	4.0	6.2											
M	315					0.12			0.27	0.44	0.62	1.0	1.6	2.5	3.9												
N	500				0.074			0.17	0.27	0.39	0.63	1.0	1.6	2.5													
P	800			0.046			0.11	0.17	0.24	0.40	0.64	0.99	1.6														
Q	1250		0.029			0.067	0.11	0.16	0.25	0.41	0.64	0.99															
R	2000	0.018			0.042	0.069	0.097	0.16	0.26	0.40	0.62																
S	3150			0.027																							

Note: For the exact AOQL, the above values must be multiplied by $\left(1 - \frac{\text{Sample size}}{\text{Lot or Batch size}}\right)$ (see 11.4)

LQ (DEFECTIVES)
10.0%

TABLE VI-A—Limiting Quality (in percent defective) for which P_a = 10 Percent (for Normal Inspection, Single sampling)

(See 11.6)

Code letter	Sample size	Acceptable Quality Level															
		0.010	0.015	0.025	0.040	0.065	0.10	0.15	0.25	0.40	0.65	1.0	1.5	2.5	4.0	6.5	10
A	2															68	
B	3														54		
C	5													37			58
D	8												25			41	54
E	13											16			27	36	44
F	20										11			18	25	30	42
G	32									6.9			12	16	20	27	34
H	50								4.5			7.6	10	13	18	22	29
J	80							2.8			4.8	6.5	8.2	11	14	19	24
K	125						1.8			3.1	4.3	5.4	7.4	9.4	12	16	23
L	200					1.2			2.0	2.7	3.3	4.6	5.9	7.7	10	14	
M	315				0.73			1.2	1.7	2.1	2.9	3.7	4.9	6.4	9.0		
N	500			0.46			0.78	1.1	1.3	1.9	2.4	3.1	4.0	5.6			
P	800		0.29			0.49	0.67	0.84	1.2	1.5	1.9	2.5	3.5				
Q	1250	0.18			0.31	0.43	0.53	0.74	0.94	1.2	1.6	2.3					
R	2000			0.20	0.27	0.33	0.46	0.59	0.77	1.0	1.4						

TABLE VI-B—Limiting Quality (in defects per hundred units) for which P_a = 10 Percent (for Normal Inspection, Single sampling)

(See 11.6)

Code letter	Sample size	Acceptable Quality Level																									
		0.010	0.015	0.025	0.040	0.065	0.10	0.15	0.25	0.40	0.65	1.0	1.5	2.5	4.0	6.5	10	15	25	40	65	100	150	250	400	650	1000
A	2															120			200	270	330	460	590	770	1000	1400	1900
B	3														77			130	180	220	310	390	510	670	940	1300	1800
C	5													46			78	110	130	190	240	310	400	560	770	1100	
D	8												29			49	67	84	120	150	190	250	350	480	670		
E	13											18			30	41	51	71	91	120	160	220	300	410			
F	20										12			20	27	33	46	59	77	100	140						
G	32									7.2			12	17	21	29	37	48	63	88							
H	50								4.6			7.8	11	13	19	24	31	40	56								
J	80							2.9			4.9	6.7	8.4	12	15	19	25	35									
K	125						1.8			3.1	4.3	5.4	7.4	9.4	12	16	23										
L	200					1.2			2.0	2.7	3.3	4.6	5.9	7.7	10	14											
M	315				0.73			1.2	1.7	2.1	2.9	3.7	4.9	6.4	9.0												
N	500			0.46			0.78	1.1	1.3	1.9	2.4	3.1	4.0	5.6													
P	800		0.29			0.49	0.67	0.84	1.2	1.5	1.9	2.5	3.5														
Q	1250	0.18			0.31	0.43	0.53	0.74	0.94	1.2	1.6	2.3															
R	2000			0.20	0.27	0.33	0.46	0.59	0.77	1.0	1.4																

LQ (DEFECTS)
10%

TABLE IX—*Average sample size curves for double and multiple sampling (normal and tightened inspection)*

(See 11.5)

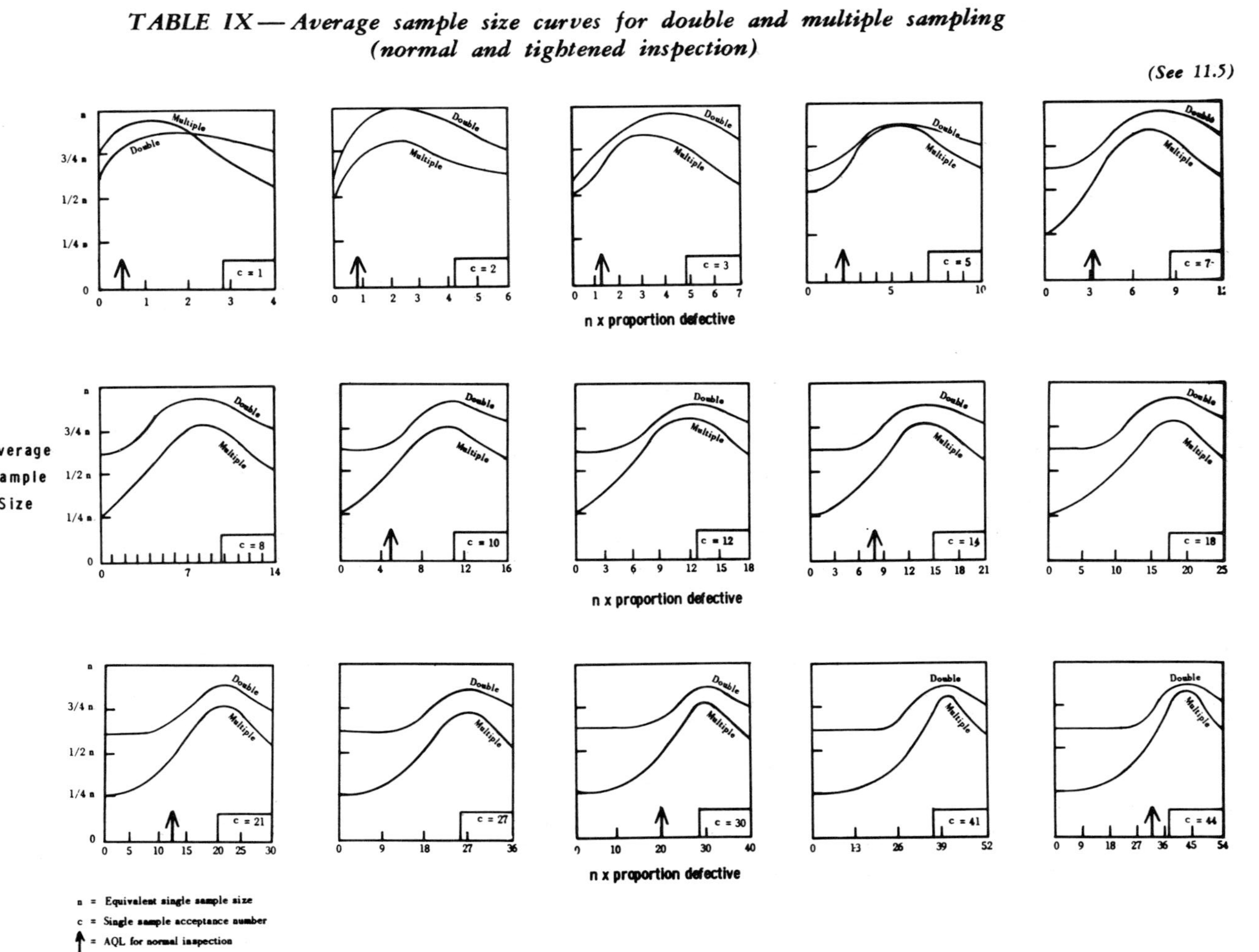

AVERAGE SAMPLE SIZE

A

TABLE X-A—Tables for sample size code letter: A

CHART A - OPERATING CHARACTERISTIC CURVES FOR SINGLE SAMPLING PLANS

(Curves for double and multiple sampling are matched as closely as practicable)

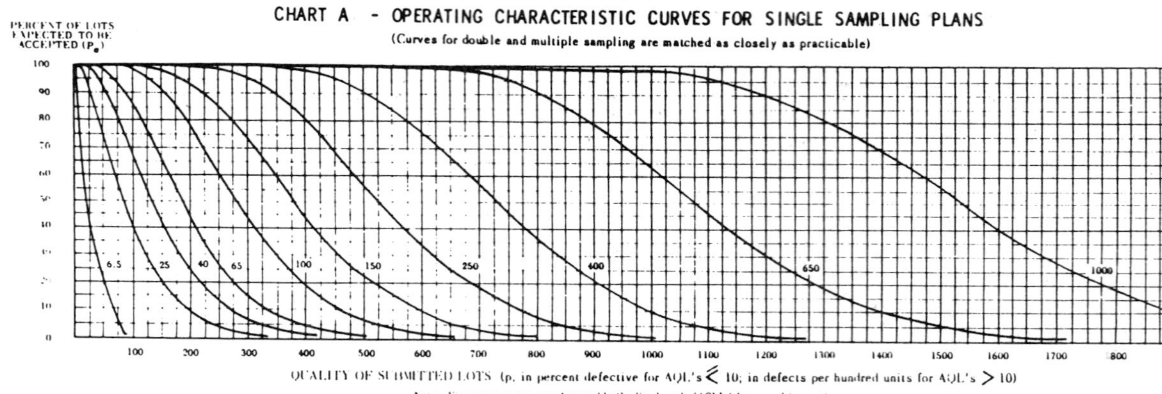

TABLE X-A-1 - TABULATED VALUES FOR OPERATING CHARACTERISTIC CURVES FOR SINGLE SAMPLING PLANS

P_a	Acceptable Quality Levels (normal inspection)														
	6.5	6.5	25	40	65	100	150	X	250	X	400	X	650	X	1000
	p (in percent defective)	p (in defects per hundred units)													
99.0	0.501	0.51	7.45	21.8	41.2	89.2	145	175	239	305	374	517	629	859	977
95.0	2.53	2.56	17.8	40.9	68.3	131	199	235	308	385	462	622	745	995	1122
90.0	5.13	5.25	26.6	55.1	87.3	158	233	272	351	432	515	684	812	1073	1206
75.0	13.4	14.4	48.1	86.8	127	211	298	342	431	521	612	795	934	1314	1354
50.0	29.3	34.7	83.9	134	184	284	383	433	533	633	733	933	1083	1383	1533
25.0	50.0	69.3	135	196	256	371	484	540	651	761	870	1087	1248	1568	1728
10.0	68.4	115	195	266	334	464	589	650	770	889	1006	1238	1409	1748	1916
5.0	77.6	150	237	315	388	526	657	722	848	972	1094	1334	1512	1862	2035
1.0	90.0	230	332	420	502	655	800	870	1007	1141	1272	1529	1718	2088	2270
	X	X	40	65	100	150	X	250	X	400	X	650	X	1000	X
	Acceptable Quality Levels (tightened inspection)														

Note: Binomial distribution used for percent defective computations; Poisson for defects per hundred units.

TABLE X-A-2 - SAMPLING PLANS FOR SAMPLE SIZE CODE LETTER: A

Type of sampling plan	Cumulative sample size	Acceptable Quality Levels (normal inspection)																		Cumulative sample size
		Less than 6.5	6.5	X	10	15	25	40	65	100	150	X	250	X	400	X	650	X	1000	
		Ac Re	Ac Re	Ac Re	Ac Re	Ac Re	Ac Re	Ac Re	Ac Re	Ac Re	Ac Re	Ac Re	Ac Re	Ac Re	Ac Re	Ac Re	Ac Re	Ac Re	Ac Re	
Single	2	▽	0 1	Use Letter D	Use Letter C	Use Letter B	1 2	2 3	3 4	5 6	7 8	8 9	10 11	12 13	14 15	18 19	21 22	27 28	30 31	2
Double		▽	•				(•)	(•)	(•)	(•)	(•)	(•)	(•)	(•)	(•)	(•)	(•)	(•)	(•)	
Multiple		▽	•				•	•	•	•	•	•	•	•	•	•	•	•	•	
		Less than 10	X	10	15	25	40	65	100	150	X	250	X	400	X	650	X	1000	X	
		Acceptable Quality Levels (tightened inspection)																		

▽ = Use next subsequent sample size code letter for which acceptance and rejection numbers are available.

Ac = Acceptance number

Re = Rejection number

• = Use single sampling plan above (or alternatively use letter D).

(•) = Use single sampling (or alternatively use letter B).

B

TABLE X-B—Tables for sample size code letter: B

CHART B - OPERATING CHARACTERISTIC CURVES FOR SINGLE SAMPLING PLANS

(Curves for double and multiple sampling are matched as closely as practicable)

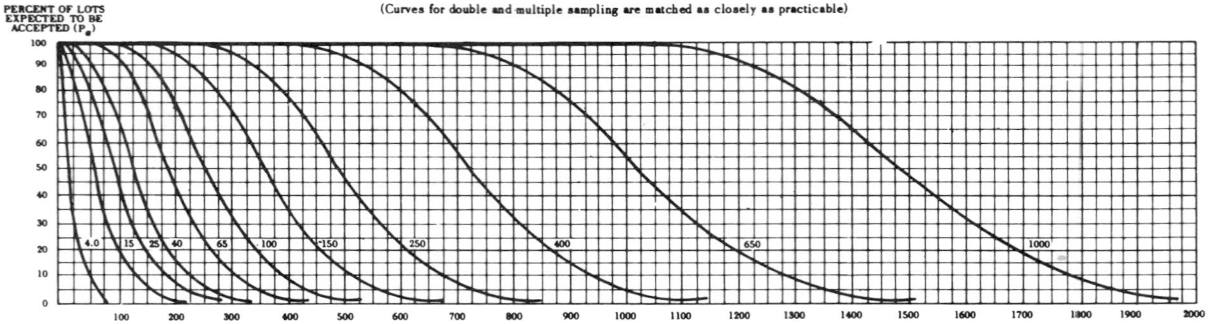

PERCENT OF LOTS EXPECTED TO BE ACCEPTED (P_a)

QUALITY OF SUBMITTED LOTS (p, in percent defective for AQL's ≤ 10; in defects per hundred units for AQL's > 10)

Note: Figures on curves are Acceptable Quality Levels (AQL's) for normal inspection.

TABLE X-B-1 - TABULATED VALUES FOR OPERATING CHARACTERISTIC CURVES FOR SINGLE SAMPLING PLANS

P_a	Acceptable Quality Levels (normal inspection)																
	4.0	4.0	15	25	40	65	100	X	150	X	250	X	400	X	650	X	1000
	p (in percent defective)	p (in defects per hundred units)															
99.0	0.33	0.34	4.97	14.5	27.4	59.5	96.9	117	159	203	249	345	419	573	651	947	1029
95.0	1.70	1.71	11.8	27.3	45.5	87.1	133	157	206	256	308	415	496	663	748	1065	1152
90.0	3.45	3.50	17.7	36.7	58.2	105	155	181	234	288	343	456	541	716	804	1131	1222
75.0	9.14	9.60	32.0	57.6	84.5	141	199	228	287	347	408	530	623	809	903	1249	1344
50.0	20.6	23.1	55.9	89.1	122	189	256	289	356	422	489	622	722	922	1022	1389	1489
25.0	37.0	46.2	89.8	131	170	247	323	360	434	507	580	724	832	1046	1152	1539	1644
10.0	53.6	76.8	130	177	223	309	392	433	514	593	671	825	939	1165	1277	1683	1793
5.0	63.2	99.9	158	210	258	350	438	481	565	648	730	890	1008	1241	1356	1773	1886
1.0	78.4	154	221	280	335	437	533	580	672	761	848	1019	1145	1392	1513	1951	2069
	6.5	6.5	25	40	65	100	X	150	X	250	X	400	X	650	X	1000	X
	Acceptable Quality Levels (tightened inspection)																

Note: Binomial distribution used for percent defective computations; Poisson for defects per hundred units.

TABLE X-B-2 - SAMPLING PLANS FOR SAMPLE SIZE CODE LETTER: B

Type of sampling plan	Cumulative sample size	Acceptable Quality Levels (normal inspection)																				Cumulative sample size
		Less than 4.0	4.0	6.5	X	10	15	25	40	65	100	X	150	X	250	X	400	X	650	X	1000	
		Ac Re	Ac Re	Ac Re	Ac Re	Ac Re	Ac Re	Ac Re	Ac Re	Ac Re	Ac Re	Ac Re	Ac Re	Ac Re	Ac Re	Ac Re	Ac Re	Ac Re	Ac Re	Ac Re	Ac Re	
Single	3	▽	0 1	Use Letter A	Use Letter D	Use Letter C	1 2	2 3	3 4	5 6	7 8	8 9	10 11	12 13	14 15	18 19	21 22	27 28	30 31	41 42	44 45	3
Double	2	▽	*				0 2	0 3	1 4	2 5	3 7	3 7	5 9	6 10	7 11	9 14	11 16	15 20	17 22	23 29	25 31	2
	4						1 2	3 4	4 5	6 7	8 9	11 12	12 13	15 16	18 19	23 24	26 27	34 35	37 38	52 53	56 57	4
Multiple		▽	*				‡	‡	‡	‡	‡	‡	‡	‡	‡	‡	‡	‡	‡	‡	‡	
		Less than 6.5	6.5	X	10	15	25	40	65	100	X	150	X	250	X	400	X	650	X	1000	X	
		Acceptable Quality Levels (tightened inspection)																				

▽ = Use next subsequent sample size code letter for which acceptance and rejection numbers are available.
Ac = Acceptance number
Re = Rejection number
• = Use single sampling plan above (or alternatively use letter E).
‡ = Use double sampling plan above (or alternatively use letter D).

B

C

TABLE X-C—Tables for sample size code letter: C

CHART C - OPERATING CHARACTERISTIC CURVES FOR SINGLE SAMPLING PLANS

(Curves for double and multiple sampling are matched as closely as practicable)

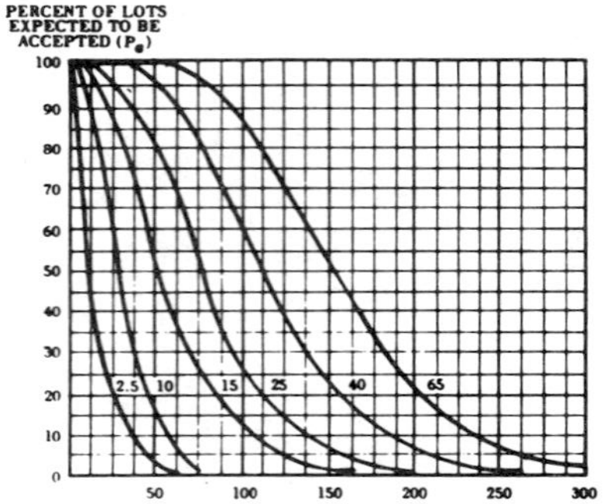

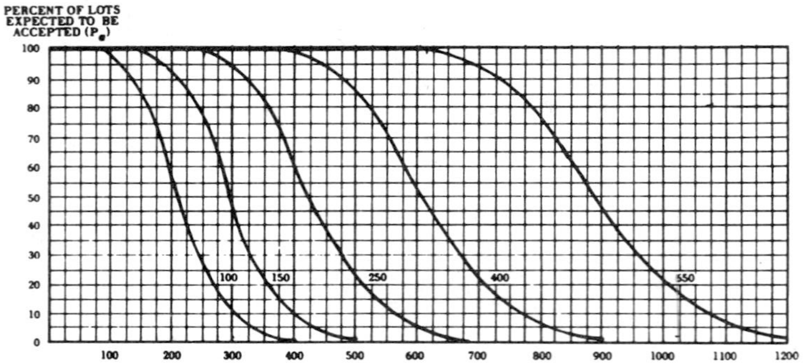

PERCENT OF LOTS EXPECTED TO BE ACCEPTED (P_a)

QUALITY OF SUBMITTED LOTS (p, in percent defective for AQL's ≤ 10; in defects per hundred units for AQL's > 10)

Note: Figures on curves are Acceptable Quality Levels (AQL's) for normal inspection.

TABLE X-C-1 - TABULATED VALUES FOR OPERATING CHARACTERISTIC CURVES FOR SINGLE SAMPLING PLANS

P_a	Acceptable Quality Levels (normal inspection)																	
	2.5	10	2.5	10	15	25	40	65	X	100	X	150	X	250	X	400	X	650
	p (in percent defective)		p (in defects per hundred units)															
99.0	0.20	3.28	0.20	2.89	8.72	16.5	35.7	58.1	70.1	95.4	122	150	207	251	344	391	568	618
95.0	1.02	7.63	1.03	7.10	16.4	27.3	52.3	79.6	93.9	123	154	185	249	298	398	449	639	691
90.0	2.09	11.2	2.10	10.6	22.0	34.9	63.0	93.1	109	140	173	206	273	325	429	482	679	733
75.0	5.59	19.4	5.76	19.2	34.5	50.7	84.4	119	137	172	208	245	318	374	485	542	749	806
50.0	12.9	31.4	13.9	33.6	53.5	73.4	113	153	173	213	253	293	373	433	553	613	833	893
25.0	24.2	45.4	27.7	53.9	78.4	102	148	194	216	260	304	348	435	499	627	691	923	987
10.0	36.9	58.4	46.1	77.8	106	134	186	235	260	308	356	403	495	564	699	766	1010	1076
5.0	45.1	65.8	59.9	94.9	126	155	210	263	289	339	389	438	534	605	745	814	1064	1131
1.0	60.2	77.8	92.1	133	168	201	262	320	348	403	456	509	612	687	835	908	1171	1241
	4.0	X	4.0	15	25	40	65	X	100	X	150	X	250	X	400	X	650	X
	Acceptable Quality Levels (tightened inspection)																	

Note: Binomial distribution used for percent defective computations; Poisson for defects per hundred units.

TABLE X-C-2 - SAMPLING PLANS FOR SAMPLE SIZE CODE LETTER: C

Type of sampling plan	Cumulative sample size	Less than 2.5	2.5	4.0	✕	6.5	10	15	25	40	65	✕	100	✕	150	✕	250	✕	400	✕	650	1000	Cumulative sample size
		Ac Re	Ac Re	Ac Re	Ac Re	Ac Re	Ac Re	Ac Re	Ac Re	Ac Re	Ac Re	Ac Re	Ac Re	Ac Re	Ac Re	Ac Re	Ac Re	Ac Re	Ac Re	Ac Re	Ac Re	Ac Re	
Single	5	▽	0 1	Use Letter B	Use Letter E	Use Letter D	1 2	2 3	3 4	5 6	7 8	8 9	10 11	12 13	14 15	18 19	21 22	27 28	30 31	41 42	44 45	Use Letter B	5
Double	3	▽	*				0 2	0 3	1 4	2 5	3 7	3 7	5 9	6 10	7 11	9 14	11 16	15 20	17 22	23 29	25 31		3
	6						1 2	3 4	4 5	6 7	8 9	11 12	12 13	15 16	18 19	23 24	26 27	34 35	37 38	52 53	56 57		6
Multiple		▽	*				++	++	++	++	++	++	++	++	++	++	++	++	++	++	++		
		Less than 4.0	4.0	✕	6.5	10	15	25	40	65	✕	100	✕	150	✕	250	✕	400	✕	650	✕	1000	
		Acceptable Quality Levels (tightened inspection)																					

▽ = Use next subsequent sample size code letter for which acceptance and rejection numbers are available.
Ac = Acceptance number.
Re = Rejection number.
* = Use single sampling plan above (or alternatively use letter F).
++ = Use double sampling plan above (or alternatively use letter D).

C

D

TABLE X-D—Tables for sample size code letter: D

CHART D - OPERATING CHARACTERISTIC CURVES FOR SINGLE SAMPLING PLANS

(Curves for double and multiple sampling are matched as closely as practicable)

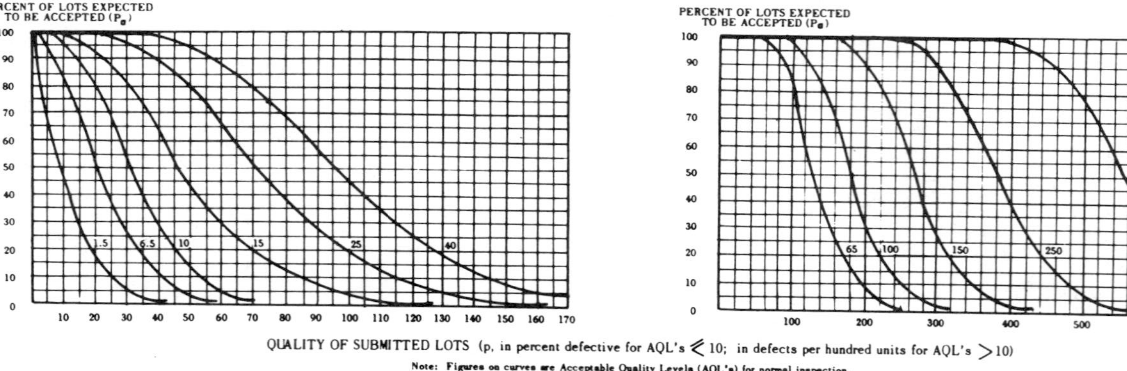

QUALITY OF SUBMITTED LOTS (p, in percent defective for AQL's ≤ 10; in defects per hundred units for AQL's > 10)

Note: Figures on curves are Acceptable Quality Levels (AQL's) for normal inspection.

TABLE X-D-1 - TABULATED VALUES FOR OPERATING CHARACTERISTIC CURVES FOR SINGLE SAMPLING PLANS

P_a	Acceptable Quality Levels (normal inspection)																		
	1.5	6.5	10	1.5	6.5	10	15	25	40	X	65	X	100	X	150	X	250	X	400
	p (in percent defective)			p (in defects per hundred units)															
99.0	0.13	2.00	6.00	0.13	1.86	5.45	10.3	22.3	36.3	43.8	59.6	76.2	93.5	129	157	215	244	355	386
95.0	0.64	2.64	11.1	0.64	4.44	10.2	17.1	32.7	49.8	58.7	77.1	96.1	116	156	186	249	281	399	432
90.0	1.31	6.88	14.7	1.31	6.65	13.8	21.8	39.4	58.2	67.9	87.8	108	129	171	203	268	301	424	458
75.0	3.53	12.1	22.1	3.60	12.0	21.6	31.7	52.7	74.5	85.5	108	130	153	199	234	303	339	468	504
50.0	8.30	20.1	32.1	8.66	21.0	33.4	45.9	70.9	95.9	108	133	158	183	233	271	346	383	521	558
25.0	15.9	30.3	43.3	17.3	33.7	49.0	63.9	92.8	121	135	163	190	218	272	312	392	432	577	617
10.0	25.0	40.6	53.9	28.8	48.6	66.5	83.5	116	147	162	193	222	252	309	352	437	478	631	672
5.0	31.2	47.1	59.9	37.5	59.3	78.7	96.9	131	164	180	212	243	274	334	378	465	509	665	707
1.0	43.8	58.8	70.7	57.6	83.0	105	126	164	200	218	252	285	318	382	429	522	568	732	776
	2.5	10	X	2.5	10	15	25	40	X	65	X	100	X	150	X	250	X	400	X
	Acceptable Quality Levels (tightened inspection)																		

TABLE X-D-2 - SAMPLING PLANS FOR SAMPLE SIZE CODE LETTER: D

Type of sampling plan	Cumulative sample size	Acceptable Quality Levels (normal inspection)																					Cumulative sample size
		Less than 1.5	1.5	2.5	X	4.0	6.5	10	15	25	40	X	65	X	100	X	150	X	250	X	400	Higher than 400	
		Ac Re	Ac Re	Ac Re	Ac Re	Ac Re	Ac Re	Ac Re	Ac Re	Ac Re	Ac Re	Ac Re	Ac Re	Ac Re	Ac Re	Ac Re	Ac Re	Ac Re	Ac Re	Ac Re	Ac Re	Ac Re	
Single	8	▽	0 1	Use Letter C	Use Letter F	Use Letter E	1 2	2 3	3 4	5 6	7 8	8 9	10 11	12 13	14 15	18 19	21 22	27 28	30 31	41 42	44 45	△	8
Double	5	▽	•				0 2	0 3	1 4	2 5	3 7	3 7	5 9	6 10	7 11	9 14	11 16	15 20	17 22	23 29	25 31	△	5
	10						1 2	3 4	4 5	6 7	8 9	11 12	12 13	15 16	18 19	23 24	26 27	34 35	37 38	52 53	56 57		10
Multiple	2	▽	•				# 2	# 2	# 3	# 4	0 4	0 4	0 5	0 6	1 7	1 8	2 9	3 10	4 12	6 15	6 16	△	2
	4						# 2	0 3	0 3	1 5	1 6	2 7	3 8	3 9	4 10	6 12	7 14	10 17	11 19	16 25	17 27		4
	6						0 2	0 3	1 4	2 6	3 8	4 9	6 10	7 12	8 13	11 17	13 19	17 24	19 27	26 36	29 39		6
	8						0 3	1 4	2 5	3 7	5 10	6 11	8 13	10 15	12 17	16 22	19 25	24 31	27 34	37 46	40 49		8
	10						1 3	2 4	3 6	5 8	7 11	9 12	11 15	14 17	17 20	22 25	25 29	32 37	36 40	49 55	53 58		10
	12						1 3	3 5	4 6	7 9	10 12	12 14	14 17	18 20	21 23	27 29	31 33	40 43	45 47	61 64	65 68		12
	14						2 3	4 5	6 7	9 10	13 14	14 15	18 19	21 22	25 26	32 33	37 38	48 49	53 54	72 73	77 78		14
		Less than 2.5	2.5	X	4.0	6.5	10	15	25	40	X	65	X	100	X	150	X	250	X	400	X	Higher than 400	
		Acceptable Quality Levels (tightened inspection)																					

△ = Use next preceding sample size code letter for which acceptance and rejection numbers are available.
▽ = Use next subsequent sample size code letter for which acceptance and rejection numbers are available.
Ac = Acceptance number
Re = Rejection number
• = Use single sampling plan above (or alternatively use letter G).
= Acceptance not permitted at this sample size.

D

E

TABLE X-E—Tables for sample size code letter: E

CHART E - OPERATING CHARACTERISTIC CURVES FOR SINGLE SAMPLING PLANS

(Curves for double and multiple sampling are matched as closely as practicable)

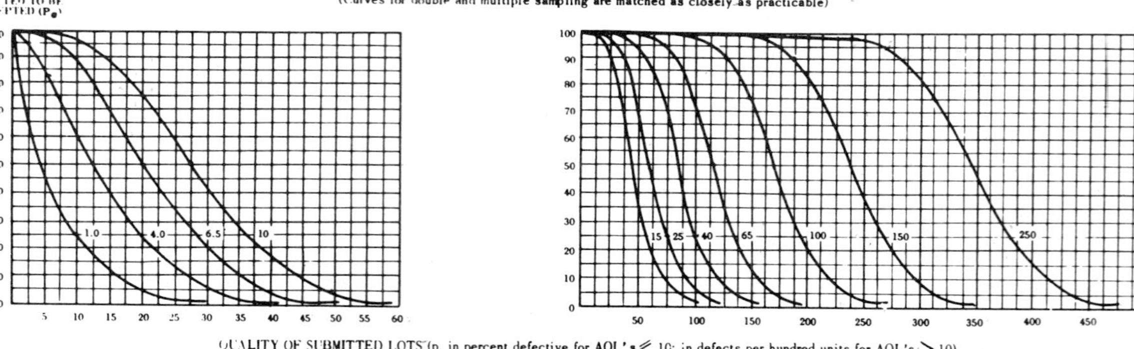

QUALITY OF SUBMITTED LOTS (p, in percent defective for AQL's ≤ 10; in defects per hundred units for AQL's > 10)

Note: Figures on curves are Acceptable Quality Levels (AQL's) for normal inspection.

TABLE X-E-1 - TABULATED VALUES FOR OPERATING CHARACTERISTIC CURVES FOR SINGLE SAMPLING PLANS

P_a	Acceptable Quality Levels (normal inspection)																			
	1.0	4.0	6.5	10	1.0	4.0	6.5	10	15	25	X	40	X	65	X	100	X	150	X	250
	p (in percent defective)				p (in defects per hundred units)															
99.0	0.077	1.19	3.63	7.00	0.078	1.15	3.35	6.33	13.7	22.4	27.0	36.7	46.9	57.5	79.6	96.7	132	150	219	238
95.0	0.394	2.81	6.63	11.3	0.395	2.73	6.29	10.5	20.1	30.6	36.1	47.5	59.2	71.1	95.7	115	153	173	246	266
90.0	0.807	4.16	8.80	14.2	0.808	4.09	8.48	13.4	24.2	35.8	41.8	54.0	66.5	79.2	105	125	165	185	261	282
75.0	2.19	7.41	13.4	19.9	2.22	7.39	13.3	19.5	32.5	45.8	52.6	66.3	80.2	94.1	122	144	187	208	288	310
50.0	5.19	12.6	20.0	27.5	5.33	12.9	20.6	28.2	43.6	59.0	66.7	82.1	97.5	113	144	168	213	236	321	344
25.0	10.1	19.4	28.0	36.2	10.7	20.7	30.2	39.3	57.1	74.5	83.1	100	117	134	167	192	241	266	355	379
10.0	16.2	26.8	36.0	44.4	17.7	29.9	40.9	51.4	71.3	90.5	100	119	137	155	190	217	269	295	388	414
5.0	20.6	31.6	41.0	49.5	23.0	36.5	48.4	59.6	80.9	101	111	130	150	168	205	233	286	313	409	435
1.0	29.8	41.5	50.6	58.7	35.4	51.1	64.7	77.3	101	123	134	155	176	196	235	264	321	349	450	477
	1.5	6.5	10	X	1.5	6.5	10	15	25	X	40	X	65	X	100	X	150	X	250	X
	Acceptable Quality Levels (tightened inspection)																			

Note: Binomial distribution used for percent defective computations; Poisson for defects per hundred units.

TABLE X-E-2 - SAMPLING PLANS FOR SAMPLE SIZE CODE LETTER: E

Type of sampling plan	Cumulative sample size	Less than 1.0	1.0	1.5	X	2.5	4.0	6.5	10	15	25	X	40	X	65	X	100	X	150	X	250	Higher than 250	Cumulative sample size
		Ac Re	Ac Re	Ac Re	Ac Re	Ac Re	Ac Re	Ac Re	Ac Re	Ac Re	Ac Re	Ac Re	Ac Re	Ac Re	Ac Re	Ac Re	Ac Re	Ac Re	Ac Re	Ac Re	Ac Re	Ac Re	
Single	13	▽	0 1	Use Letter D	Use Letter G	Use Letter F	1 2	2 3	3 4	5 6	7 8	8 9	10 11	12 13	14 15	18 19	21 22	27 28	30 31	41 42	44 45	△	13
Double	8	▽	•				0 2	0 3	1 4	2 5	3 7	3 7	5 9	6 10	7 11	9 14	11 16	15 20	17 22	23 29	25 31	△	8
	16						1 2	3 4	4 5	6 7	8 9	11 12	12 13	15 16	18 19	23 24	26 27	34 35	37 38	52 53	56 57		16
Multiple	3	▽	•				# 2	# 2	# 3	# 4	0 4	0 4	0 5	0 6	1 7	1 8	2 9	3 10	4 12	6 15	6 16	△	3
	6						# 2	0 3	0 3	1 5	1 6	2 7	3 8	3 9	4 10	6 12	7 14	10 17	11 19	16 25	17 27		6
	9						0 2	0 3	1 4	2 6	3 8	4 9	6 10	7 12	8 13	11 17	13 19	17 24	19 27	26 36	29 39		9
	12						0 3	1 4	2 5	3 7	5 10	6 11	8 13	10 15	12 17	16 22	19 25	24 31	27 34	37 46	40 49		12
	15						1 3	2 4	3 6	5 8	7 11	9 12	11 15	14 17	17 20	22 25	25 29	32 37	36 40	49 55	53 58		15
	18						1 3	3 5	4 6	7 9	10 12	12 14	14 17	18 20	21 23	27 29	31 33	40 43	45 47	61 64	65 68		18
	21						2 3	4 5	6 7	9 10	13 14	14 15	18 19	21 22	25 26	32 33	37 38	48 49	53 54	72 73	77 78		21
Acceptable Quality Levels (tightened inspection)		Less than 1.5	1.5	X	2.5	4.0	6.5	10	15	25	X	40	X	65	X	100	X	150	X	250	X	Higher than 250	

Columns headed "Less than 1.0" through "Higher than 250": Acceptable Quality Levels (normal inspection).

△ = Use next preceding sample size code letter for which acceptance and rejection numbers are available.

▽ = Use next subsequent sample size code letter for which acceptance and rejection numbers are available.

Ac = Acceptance number.

Re = Rejection number.

• = Use single sampling plan above (or alternatively use letter H).

= Acceptance not permitted at this sample size.

E

F

TABLE X-F—Tables for sample size code letter: F

CHART F - OPERATING CHARACTERISTIC CURVES FOR SINGLE SAMPLING PLANS

(Curves for double and multiple sampling are matched as closely as practicable)

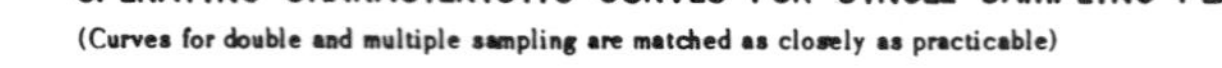
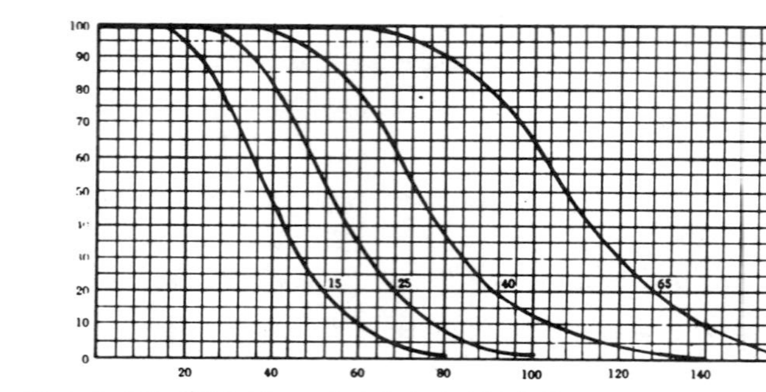

QUALITY OF SUBMITTED LOTS (p, in percent defective for AQL's ≤10; in defects per hundred units for AQL's >10)

Note: Figures on curves are Acceptable Quality Levels (AQL's) for normal inspection.

TABLE X-F-1 - TABULATED VALUES FOR OPERATING CHARACTERISTIC CURVES FOR SINGLE SAMPLING PLANS

P_a	Acceptable Quality Levels (normal inspection)																
	0.65	2.5	4.0	6.5	10	0.65	2.5	4.0	6.5	10	15	X	25	X	40	X	65
	p (in percent defective)					p (in defects per hundred units)											
99.0	0.050	0.75	2.25	4.31	9.75	0.051	0.75	2.18	4.12	8.92	14.5	17.5	23.9	30.5	37.4	51.7	62.9
95.0	0.256	1.80	4.22	7.13	14.0	0.257	1.78	4.09	6.83	13.1	19.9	23.5	30.8	38.5	46.2	62.2	74.5
90.0	0.525	2.69	5.64	9.03	16.6	0.527	2.66	5.51	8.73	15.8	23.3	27.2	35.1	43.2	51.5	68.4	81.2
75.0	1.43	4.81	8.70	12.8	21.6	1.44	4.81	8.68	12.7	21.1	29.8	34.2	43.1	52.1	61.2	79.5	93.4
50.0	3.41	8.25	13.1	18.1	27.9	3.47	8.39	13.4	18.4	28.4	38.3	43.3	53.3	63.3	73.3	93.3	108
25.0	6.70	12.9	18.7	24.2	34.8	6.93	13.5	19.6	25.5	37.1	48.4	54.0	65.1	76.1	87.0	109	125
10.0	10.9	18.1	24.5	30.4	41.5	11.5	19.5	26.6	33.4	46.4	58.9	65.0	77.0	88.9	101	124	141
5.0	13.9	21.6	28.3	34.4	45.6	15.0	23.7	31.5	38.8	52.6	65.7	72.2	84.8	97.2	109	133	151
1.0	20.6	28.9	35.6	42.0	53.4	23.0	33.2	42.0	50.2	65.5	80.0	87.0	101	114	127	153	172
	1.0	4.0	6.5	10	X	1.0	4.0	6.5	10	15	X	25	X	40	X	65	X
	Acceptable Quality Levels (tightened inspection)																

Note: Binomial distribution used for percent defective computations; Poisson for defects per hundred units.

TABLE X-F-2 - SAMPLING PLANS FOR SAMPLE SIZE CODE LETTER: F

Type of sampling plan	Cumulative sample size	Acceptable Quality Levels (normal inspection)																																		Cumulative sample size
		Less than 0.65		0.65		1.0		X		1.5		2.5		4.0		6.5		10		15		X		25		X		40		X		65		Higher than 65		
		Ac	Re	Ac	Re	Ac	Re	Ac	He	Ac	Re	Ac	Re	Ac	Re	Ac	Re	Ac	Re	Ac	Re	Ac	Re	Ac	Re	Ac	Re	Ac	Re	Ac	Re	Ac	Re	Ac	Re	
Single	20	▽		0	1	Use Letter E		Use Letter H		Use Letter G		1	2	2	3	3	4	5	6	7	8	8	9	10	11	12	13	14	15	18	19	21	22	△		20
Double	13	▽		•								0	2	0	3	1	4	2	5	3	7	3	7	5	9	6	10	7	11	9	14	11	16	△		13
	26											1	2	3	4	4	5	6	7	8	9	11	12	12	13	15	16	18	19	23	24	26	27			26
Multiple	5	▽		•								#	2	#	2	#	3	#	4	0	4	0	4	0	5	0	6	1	7	1	8	2	9	△		5
	10											#	2	0	3	0	3	1	5	1	6	2	7	3	8	3	9	4	10	6	12	7	14			10
	15											0	2	0	3	1	4	2	6	3	8	4	9	6	10	7	12	8	13	11	17	13	19			15
	20											0	3	1	4	2	5	3	7	5	10	6	11	8	13	10	15	12	17	16	22	19	25			20
	25											1	3	2	4	3	6	5	8	7	11	9	12	11	15	14	17	17	20	22	25	25	29			25
	30											1	3	3	5	4	6	7	9	10	12	12	14	14	17	18	20	21	23	27	29	31	33			30
	35											2	3	4	5	6	7	9	10	13	14	14	15	18	19	21	22	25	26	32	33	37	38			35
		Less than 1.0		1.0		X		1.5		2.5		4.0		6.5		10		15		X		25		X		40		X		65		X		Higher than 65		
		Acceptable Quality Levels (tightened inspection)																																		

△ = Use next preceding sample size code letter for which acceptance and rejection numbers are available.
▽ = Use next subsequent sample size code letter for which acceptance and rejection numbers are available.
Ac = Acceptance number
He = Rejection number
• = Use single sampling plan above (or alternatively use letter J).
= Acceptance not permitted at this sample size.

F

G

TABLE X-G—Tables for sample size code letter: G

CHART G - OPERATING CHARACTERISTIC CURVES FOR SINGLE SAMPLING PLANS

(Curves for double and multiple sampling are matched as closely as practicable)

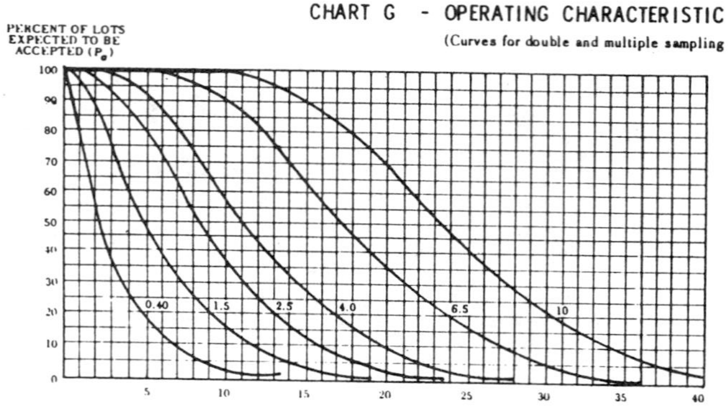

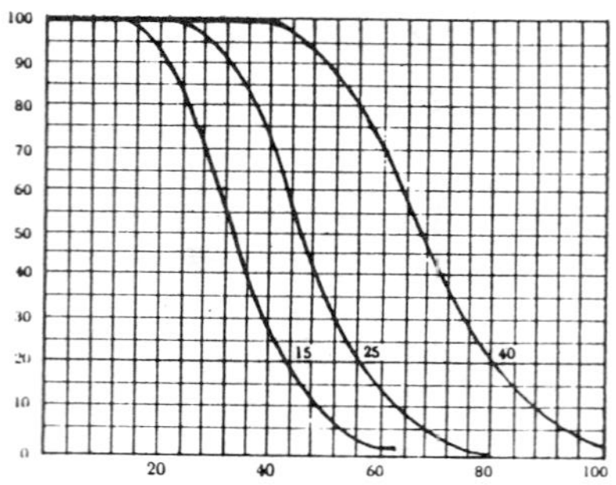

PERCENT OF LOTS EXPECTED TO BE ACCEPTED (P_a)

QUALITY OF SUBMITTED LOTS (p, in percent defective for AQL's ⩽ 10; in defects per hundred units for AQL's > 10)

Note: Figures on curves are Acceptable Quality Levels (AQL's) for normal inspection.

TABLE X-G-1 - TABULATED VALUES FOR OPERATING CHARACTERISTIC CURVES FOR SINGLE SAMPLING PLANS

P_a	Acceptable Quality Levels (normal inspection)																	
	0.40	1.5	2.5	4.0	6.5	10	0.40	1.5	2.5	4.0	6.5	10	X	15	X	25	X	40
	p (in percent defective)						p (in defects per hundred units)											
99.0	0.032	0.475	1.38	2.63	5.94	9.75	0.032	0.466	1.36	2.57	5.57	9.08	11.0	14.9	19.1	23.4	32.3	39.3
95.0	0.161	1.13	2.59	4.39	8.50	13.1	0.160	1.10	2.55	4.26	8.16	12.4	14.7	19.3	24.0	28.9	38.9	46.5
90.0	0.329	1.67	3.50	5.56	10.2	15.1	0.328	1.66	3.44	5.45	9.85	14.6	17.0	21.9	27.0	32.2	42.7	50.8
75.0	0.895	3.01	5.42	7.98	13.4	19.0	0.900	3.00	5.39	7.92	13.2	18.6	21.4	26.9	32.6	38.2	49.7	58.4
50.0	2.14	5.19	8.27	11.4	17.5	23.7	2.16	5.24	8.35	11.5	17.7	24.0	27.1	33.3	39.6	45.8	58.3	67.7
25.0	4.23	8.19	11.9	15.4	22.3	29.0	4.33	8.41	12.3	16.0	23.2	30.3	33.8	40.7	47.6	54.4	67.9	78.0
10.0	6.94	11.6	15.8	19.7	27.1	34.1	7.19	12.2	16.6	20.9	29.0	36.8	40.6	48.1	55.6	62.9	77.4	88.1
5.0	8.94	14.0	18.4	22.5	30.1	37.2	9.36	14.8	19.7	24.2	32.9	41.1	45.1	53.0	60.8	68.4	83.4	94.5
1.0	13.5	19.0	23.7	28.0	35.9	43.3	14.4	20.7	26.3	31.4	41.0	50.0	54.4	63.0	71.3	79.5	95.5	107
	0.65	2.5	4.0	6.5	10	X	0.65	2.5	4.0	6.5	10	X	15	X	25	X	40	X
	Acceptable Quality Levels (tightened inspection)																	

Note: Binomial distribution used for percent defective computations; Poisson for defects per hundred units.

TABLE X-G-2 - SAMPLING PLANS FOR SAMPLE SIZE CODE LETTER: G

Type of sampling plan	Cumulative sample size	Acceptable Quality Levels (normal inspection)																																		Cumulative sample size
		Less than 0.40		0.40		0.65		X		1.0		1.5		2.5		4.0		6.5		10		X		15		X		25		X		40		Higher than 40		
		Ac	Re	Ac	Re	Ac	Re	Ac	Re	Ac	Re	Ac	Re	Ac	Re	Ac	Re	Ac	Re	Ac	Re	Ac	Re	Ac	Re	Ac	Re	Ac	Re	Ac	Re	Ac	Re	Ac	Re	
Single	32	▽		0	1	Use Letter F		Use Letter J		Use Letter H		1	2	2	3	3	4	5	6	7	8	8	9	10	11	12	13	14	15	18	19	21	22	△		32
Double	20	▽		*								0	2	0	3	1	4	2	5	3	7	3	7	5	9		10	7	11	9	14	11	16	△		20
	40											1	2	3	4	4	5	6	7	8	9	11	12	12	13	15	16	18	19	23	24	26	27			40
Multiple	8	▽		*								#	2	#	2	#	3	#	4	0	4	0	4	0	5	0	6	1	7	1	8	2	9	△		8
	16											#	2	0	3	0	3	1	5	1	6	2	7	3	8	3	9	4	10	6	12	7	14			16
	24											0	2	0	3	1	4	2	6	3	8	4	9	6	10	7	12	8	13	11	17	13	19			24
	32											0	3	1	4	2	5	3	7	5	10	6	11	8	13	10	15	12	17	16	22	19	25			32
	40											1	3	2	4	3	6	5	8	7	11	9	12	11	15	14	17	17	20	22	25	25	29			40
	48											1	3	3	5	4	6	7	9	10	12	12	14	14	17	18	20	21	23	27	29	31	33			48
	56											2	3	4	5	6	7	9	10	13	14	14	15	18	19	21	22	25	26	32	33	37	38			56
		Less than 0.65		0.65		X		1.0		1.5		2.5		4.0		6.5		10		X		15		X		25		X		40		X		Higher than 40		
		Acceptable Quality Levels (tightened inspection)																																		

△ = Use next preceding sample size code letter for which acceptance and rejection numbers are available.
▽ = Use next subsequent sample size code letter for which acceptance and rejection numbers are available.
Ac = Acceptance number.
Re = Rejection number.
* = Use single sampling plan above (or alternatively use letter K).
\# = Acceptance not permitted at this sample size.

G

H

TABLE X-H—Tables for sample size code letter: H

CHART H - OPERATING CHARACTERISTIC CURVES FOR SINGLE SAMPLING PLANS

(Curves for double and multiple sampling are matched as closely as practicable)

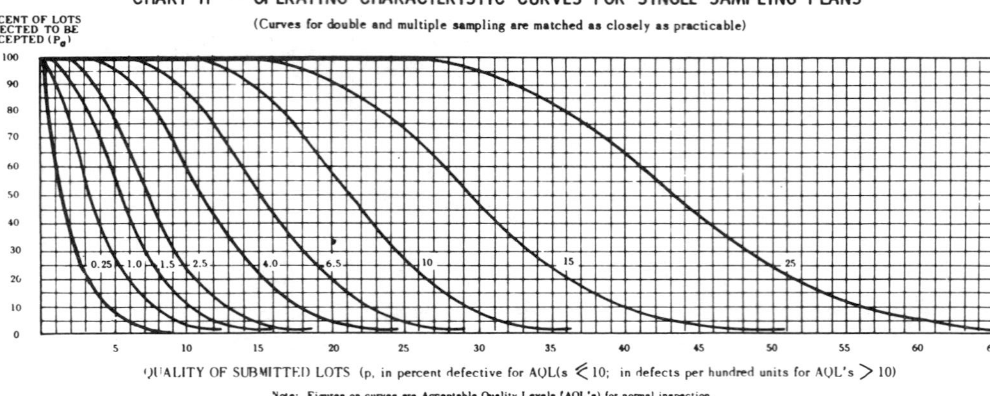

QUALITY OF SUBMITTED LOTS (p, in percent defective for AQL(s ≤ 10; in defects per hundred units for AQL's > 10)

Note: Figures on curves are Acceptable Quality Levels (AQL's) for normal inspection.

TABLE X-H-1 - TABULATED VALUES FOR OPERATING CHARACTERISTIC CURVES FOR SINGLE SAMPLING PLANS

P_a	Acceptable Quality Levels (normal inspection)																			
	0.25	1.0	1.5	2.5	4.0	6.5	X	10	0.25	1.0	1.5	2.5	4.0	6.5	X	10	X	15	X	25
	p (in percent defective)								p (in defects per hundred units)											
99.0	0.020	0.306	0.888	1.69	3.66	6.06	7.41	11.1	0.020	0.298	0.872	1.65	3.57	5.81	7.01	9.54	12.2	15.0	20.7	25.1
95.0	0.103	0.712	1.66	2.77	5.34	8.20	9.74	12.9	0.103	0.710	1.64	2.73	5.23	7.96	9.39	12.3	15.4	18.5	24.9	29.8
90.0	0.210	1.07	2.23	3.54	6.42	9.53	11.2	14.5	0.210	1.06	2.20	3.49	6.30	9.31	10.9	14.0	17.3	20.6	27.3	32.5
75.0	0.574	1.92	3.46	5.09	8.51	12.0	13.8	17.5	0.576	1.92	3.45	5.07	8.44	11.9	13.7	17.2	20.8	24.5	31.8	37.4
50.0	1.38	3.33	5.31	7.30	11.3	15.2	17.2	21.2	1.39	3.36	5.35	7.34	11.3	15.3	17.3	21.6	25.3	29.3	37.3	43.3
25.0	2.74	5.30	7.70	10.0	14.5	18.8	21.0	25.2	2.77	5.39	7.84	10.2	14.8	19.4	21.6	26.0	30.4	34.8	43.5	49.9
10.0	4.50	7.56	10.3	12.9	17.8	22.4	24.7	29.1	4.61	7.78	10.6	13.4	18.6	23.5	26.0	30.8	35.6	40.3	49.5	56.4
5.0	5.82	9.13	12.1	14.8	19.9	24.7	27.0	31.6	5.99	9.49	12.6	15.5	21.0	26.3	28.9	33.9	38.9	43.8	53.4	60.5
1.0	8.80	12.5	15.9	18.8	24.3	29.2	31.7	36.3	9.21	13.3	16.8	20.1	26.2	32.0	34.8	40.3	45.6	50.9	61.1	68.7
	0.40	1.5	2.5	4.0	6.5	X	10	X	0.40	1.5	2.5	4.0	6.5	X	10	X	15	X	25	X
	Acceptable Quality Levels (tightened inspection)																			

Note: Binomial distribution used for percent defective computations; Poisson for defects per hundred units.

TABLE X-H-2 - SAMPLING PLANS FOR SAMPLE SIZE CODE LETTER: H

Type of sampling plan	Cumulative sample size	Acceptable Quality Levels (normal inspection)																																		Cumulative sample size
		Less than 0.25		0.25		0.40		X		0.65		1.0		1.5		2.5		4.0		6.5		X		10		X		15		X		25		Higher than 25		
		Ac	Re	Ac	Re	Ac	Re	Ac	Re	Ac	Re	Ac	Re	Ac	Re	Ac	Re	Ac	Re	Ac	Re	Ac	Re	Ac	Re	Ac	Re	Ac	Re	Ac	Re	Ac	Re	Ac	Re	
Single	50	▽		0	1	Use Letter G		Use Letter K		Use Letter J		1	2	2	3	3	4	5	6	7	8	8	9	10	11	12	13	14	15	18	19	21	22	△		50
Double	32	▽		•								0	2	0	3	1	4	2	5	3	7	3	7	5	9	6	10	7	11	9	14	11	16	△		32
	64											1	2	3	4	4	5	6	7	8	9	11	12	12	13	15	16	18	19	23	24	26	27			64
Multiple	13	▽		•								#	2	#	2	#	3	#	4	0	4	0	4	0	5	0	6	1	7	1	8	2	9	△		13
	26											#	2	0	3	0	3	1	5	1	6	2	7	3	8	3	9	4	10	6	12	7	14			26
	39											0	2	0	3	1	4	2	6	3	8	4	9	6	10	7	12	8	13	11	17	13	19			39
	52											0	3	1	4	2	5	3	7	5	10	6	11	8	13	10	15	12	17	16	22	19	25			52
	65											1	3	2	4	3	6	5	8	7	11	9	12	11	15	14	17	17	20	22	25	25	29			65
	78											1	3	3	5	4	6	7	9	10	12	12	14	14	17	18	20	21	23	27	29	31	33			78
	91											2	3	4	5	6	7	9	10	13	14	14	15	18	19	21	22	25	26	32	33	37	38			91
		Less than 0.40		0.40		X		0.65		1.0		1.5		2.5		4.0		6.5		X		10		X		15		X		25		X		Higher than 25		
		Acceptable Quality Levels (tightened inspection)																																		

△ = Use next preceding sample size code letter for which acceptance and rejection numbers are available.

▽ = Use next subsequent sample size code letter for which acceptance and rejection numbers are available.

Ac = Acceptance number

Re = Rejection number

• = Use single sampling plan above (or alternatively use letter L).

= Acceptance not permitted at this sample size.

J

TABLE X-J—Tables for sample size code letter: J

CHART J - OPERATING CHARACTERISTIC CURVES FOR SINGLE SAMPLING PLANS

(Curves for double and multiple sampling are matched as closely as practicable)

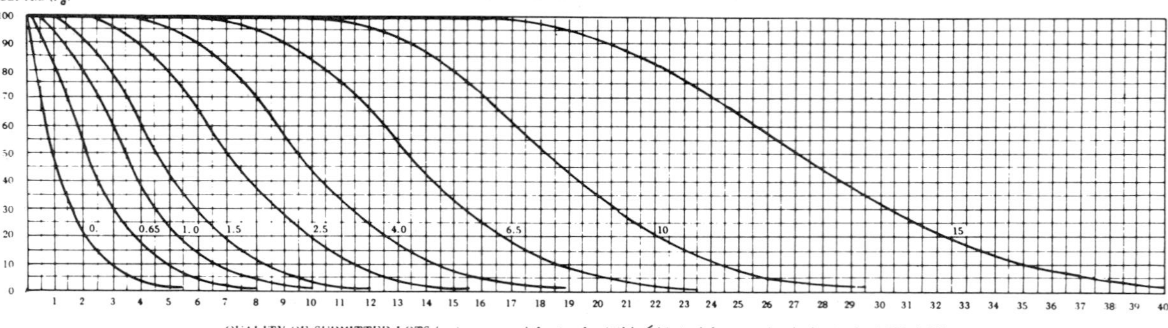

QUALITY OF SUBMITTED LOTS (p, in percent defective for AQL's ≤10; in defects per hundred units for AQL's >10)

Note: Figures on curves are Acceptable Quality Levels (AQL's) for normal inspection.

TABLE X-J-1 - TABULATED VALUES FOR OPERATING CHARACTERISTIC CURVES FOR SINGLE SAMPLING PLANS

P_a	Acceptable Quality Levels (normal inspection)																					
	0.15	0.65	1.0	1.5	2.5	4.0	X	6.5	X	10	0.15	0.65	1.0	1.5	2.5	4.0	X	6.5	X	10	X	15
	p (in percent defective)										p (in defects per hundred units)											
99.0	0.013	0.188	0.550	1.05	2.30	3.72	4.50	6.13	7.88	9.75	0.013	0.186	0.545	1.03	2.23	3.63	4.38	5.96	7.62	9.35	12.9	15.7
95.0	0.064	0.444	1.03	1.73	3.32	5.06	5.98	7.91	9.89	11.9	0.064	0.444	1.02	1.71	3.27	4.98	5.87	7.71	9.61	11.6	15.6	18.6
90.0	0.132	0.666	1.38	2.20	3.98	5.91	6.91	8.95	11.0	13.2	0.131	0.665	1.38	2.18	3.94	5.82	6.79	8.78	10.8	12.9	17.1	20.3
75.0	0.359	1.202	2.16	3.18	5.30	7.50	8.62	10.9	13.2	15.5	0.360	1.20	2.16	3.17	5.27	7.45	8.55	10.8	13.0	15.3	19.9	23.4
50.0	0.863	2.09	3.33	4.57	7.06	9.55	10.8	13.3	15.8	18.3	0.866	2.10	3.34	4.59	7.09	9.59	10.8	13.3	15.8	18.3	23.3	27.1
25.0	1.72	3.33	4.84	6.31	9.14	11.9	13.3	16.0	18.6	21.3	1.73	3.37	4.90	6.39	9.28	12.1	13.5	16.3	19.0	21.8	27.2	31.2
10.0	2.84	4.78	6.52	8.16	11.3	14.2	15.7	18.6	21.4	24.2	2.88	4.86	6.65	8.35	11.6	14.7	16.2	19.3	22.2	25.2	30.9	35.2
5.0	3.68	5.80	7.66	9.39	12.7	15.8	17.3	20.3	23.2	26.0	3.75	5.93	7.87	9.69	13.1	16.4	18.0	21.2	24.3	27.4	33.4	37.8
1.0	5.59	8.00	10.1	12.0	15.6	18.9	20.5	23.6	26.5	29.5	5.76	8.30	10.5	12.6	16.4	20.0	21.8	25.2	28.5	31.8	38.2	42.9
	0.25	1.0	1.5	2.5	4.0	X	6.5	X	10	X	0.25	1.0	1.5	2.5	4.0	X	6.5	X	10	X	15	X
	Acceptable Quality Levels (tightened inspection)																					

Note: All values given in above table based on Poisson distribution as an approximation to the Binomial.

TABLE X-J-2 - SAMPLING PLANS FOR SAMPLE SIZE CODE LETTER: J

Type of sampling plan	Cumulative sample size	Acceptable Quality Levels (normal inspection)																																		Cumulative sample size
		Less than 0.15		0.15		0.25		X		0.40		0.65		1.0		1.5		2.5		4.0		X		6.5		X		10		X		15		Higher than 15		
		Ac	Re	Ac	Re	Ac	Rc	Ac	Re	Ac	Re	Ac	Re	Ac	Re	Ac	Re	Ac	Re	Ac	Re	Ac	Re	Ac	Re	Ac	Re	Ac	Re	Ac	Re	Ac	Re	Ac	Re	
Single	80	▽		0	1	Use Letter H		Use Letter L		Use Letter K		1	2	2	3	3	4	5	6	7	8	8	9	10	11	12	13	14	15	18	19	21	22	△		80
Double	50	▽		•								0	2	0	3	1	4	2	5	3	7	3	7	5	9	6	10	7	11	9	14	11	16	△		50
	100											1	2	3	4	4	5	6	7	8	9	11	12	12	13	15	16	18	19	23	24	26	27			100
Multiple	20	▽		•								#	2	#	2	#	3	#	4	0	4	0	4	0	5	0	6	1	7	1	8	2	9	△		20
	40											#	2	0	3	0	3	1	5	1	6	2	7	3	8	3	9	4	10	6	12	7	14			40
	60											0	2	0	3	1	4	2	6	3	8	4	9	6	10	7	12	ε	13	11	17	13	19			60
	80											0	3	1	4	2	5	3	7	5	10	6	11	8	13	10	15	12	17	16	22	19	25			80
	100											1	3	2	4	3	6	5	8	7	11	9	12	11	15	14	17	17	20	22	25	25	29			100
	120											1	3	3	5	4	6	7	9	10	12	12	14	14	17	18	20	21	23	27	29	31	33			120
	140											2	3	4	5	6	7	9	10	13	14	14	15	18	19	21	22	25	26	32	33	37	38			140
		Less than 0.25		0.25		X		0.40		0.65		1.0		1.5		2.5		4.0		X		6.5		X		10		X		15		X		Higher than 15		
		Acceptable Quality Levels (tightened inspection)																																		

△ = Use next preceding sample size code letter for which acceptance and rejection numbers are available.

▽ = Use next subsequent sample size code letter for which acceptance and rejection numbers are available.

Ac = Acceptance number

Re = Rejection number

• = Use single sampling plan above (or alternatively use letter M)

\# = Acceptance not permitted at this sample size.

J

K

TABLE X-K—*Tables for sample size code letter:* K

CHART K - OPERATING CHARACTERISTIC CURVES FOR SINGLE SAMPLING PLANS

(Curves for double and multiple sampling are matched as closely as practicable)

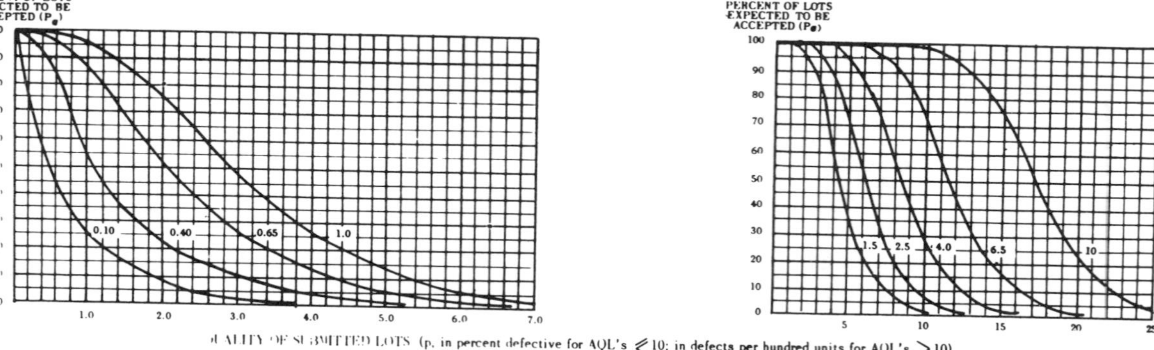

QUALITY OF SUBMITTED LOTS (p, in percent defective for AQL's ≤10; in defects per hundred units for AQL's > 10)

Note: Figures on curves are Acceptable Quality Levels (AQL's) for normal inspection.

TABLE X-K-1 - TABULATED VALUES FOR OPERATING CHARACTERISTIC CURVES FOR SINGLE SAMPLING PLANS

P_a	Acceptable Quality Levels (normal inspection)											
	0.10	0.40	0.65	1.0	1.5	2.5	X	4.0	X	6.5	X	10
	p (in percent defective or defects per hundred units)											
99.0	0.0081	0.119	0.349	0.658	1.43	2.33	2.81	3.82	4.88	5.98	8.28	10.1
95.0	0.0410	0.284	0.654	1.09	2.09	3.19	3.76	4.94	6.15	7.40	9.95	11.9
90.0	0.0840	0.426	0.882	1.40	2.52	3.73	4.35	5.62	6.92	8.24	10.9	13.0
75.0	0.230	0.769	0.382	2.03	3.38	4.77	5.47	6.90	8.34	9.79	12.7	14.9
50.0	0.554	1.34	2.14	2.94	4.54	6.14	6.94	8.53	10.1	11.7	14.9	17.3
25.0	1.11	2.15	3.14	4.09	5.94	7.75	8.64	10.4	12.2	13.9	17.4	20.0
10.0	1.84	3.11	4.26	5.35	7.42	9.42	10.4	12.3	14.2	16.1	19.8	22.5
5.0	2.40	3.80	5.04	6.20	8.41	10.5	11.5	13.6	15.6	17.5	21.4	24.2
1.0	3.68	5.31	6.73	8.04	10.5	12.8	18.3	16.1	18.3	20.4	24.5	27.5
	0.15	0.65	1.0	1.5	2.5	X	4.0	X	6.5	X	10	X
	Acceptable Quality Levels (tightened inspection)											

Note: All values given in above table based on Poisson distribution as an approximation to the Binomial.

TABLE X-K-2 - SAMPLING PLANS FOR SAMPLE SIZE CODE LETTER: K

Type of sampling plan	Cumu-lative sample size	Acceptable Quality Levels (normal inspection)																																		Cumu-lative sample size
		Less than 0.10		0.10		0.15		×		0.25		0.40		0.65		1.0		1.5		2.5		×		4.0		×		6.5		×		10		Higher than 10		
		Ac	Re	Ac	Re	Ac	Re	Ac	Re	Ac	Re	Ac	Re	Ac	Re	Ac	Re	Ac	Re	Ac	Re	Ac	Re	Ac	Re	Ac	Re	Ac	Re	Ac	Re	Ac	Re	Ac	Re	
Single	125	▽		0	1	Use Letter J		Use Letter M		Use Letter L		1	2	2	3	3	4	5	6	7	8	8	9	10	11	12	13	14	15	18	19	21	22	△		125
Double	80	▽		*								0	2	0	3	1	4	2	5	3	7	3	7	5	9	6	10	7	11	9	14	11	16	△		80
	160											1	2	3	4	4	5	6	7	8	9	11	12	12	13	15	16	18	19	23	24	26	27			160
Multiple	32	▽		*								#	2	#	2	#	3	#	4	0	4	0	4	0	5	0	6	1	7	1	8	2	9	△		32
	64											#	2	0	3	0	3	1	5	1	6	2	7	3	8	3	9	4	10	6	12	7	14			64
	96											0	2	0	3	1	4	2	6	3	8	4	9	6	10	7	12	8	13	11	17	13	19			96
	128											0	3	1	4	2	5	3	7	5	10	6	11	8	13	10	15	12	17	16	22	19	25			128
	160											1	3	2	4	3	6	5	8	7	11	9	12	11	15	14	17	17	20	22	25	25	29			160
	192											1	3	3	5	4	6	7	9	10	12	12	14	14	17	18	20	21	23	27	29	31	33			192
	224											2	3	4	5	6	7	9	10	13	14	14	15	18	19	21	22	25	26	32	33	37	38			224
		Less than 0.15		0.15		×		0.25		0.40		0.65		1.0		1.5		2.5		×		4.0		×		6.5		×		10		×		Higher than 10		
		Acceptable Quality Levels (tightened inspection)																																		

△ = Use next preceding sample size code letter for which acceptance and rejection numbers are available.
▽ = Use next subsequent sample size code letter for which acceptance and rejection numbers are available.
Ac = Acceptance number
Re = Rejection number
* = Use single sampling plan above (or alternatively use letter N).
# = Acceptance not permitted at this sample size.

K

L

TABLE X-L—Tables for sample size code letter: L

CHART L - OPERATING CHARACTERISTIC CURVES FOR SINGLE SAMPLING PLANS

(Curves for double and multiple sampling are matched as closely as practicable)

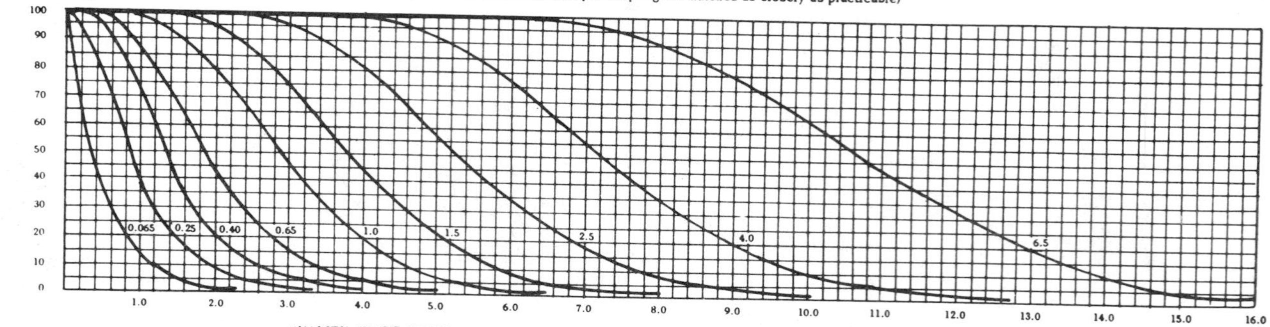

PERCENT OF LOTS EXPECTED TO BE ACCEPTED (P_a)

100 90 80 70 60 50 40 30 20 10 0

0.065 0.25 0.40 0.65 1.0 1.5 2.5 4.0 6.5

1.0 2.0 3.0 4.0 5.0 6.0 7.0 8.0 9.0 10.0 11.0 12.0 13.0 14.0 15.0 16.0

QUALITY OF SUBMITTED LOTS (p, in percent defective for AQL's ≤ 10; in defects per hundred units for AQL's > 10)

Note: Figures on curves are Acceptable Quality Levels (AQL's) for normal inspection.

TABLE X-L-1 - TABULATED VALUES FOR OPERATING CHARACTERISTIC CURVES FOR SINGLE SAMPLING PLANS

P_a	Acceptable Quality Levels (normal inspection)											
	0.065	0.25	0.40	0.65	1.0	1.5	X	2.5	X	4.0	X	6.5
	p (in percent defective or defects per hundred units)											
99.0	0.0051	0.075	0.218	0.412	0.893	1.45	1.75	2.39	3.05	3.74	5.17	6.29
95.0	0.0256	0.178	0.409	0.683	1.31	1.99	2.35	3.09	3.85	4.62	6.22	7.45
90.0	0.0525	0.266	0.551	0.873	1.58	2.33	2.72	3.51	4.32	5.15	6.84	8.12
75.0	0.144	0.481	0.864	1.27	2.11	2.98	3.42	4.31	5.21	6.12	7.95	9.34
50.0	0.347	0.839	1.34	1.84	2.84	3.84	4.33	5.33	6.33	7.33	9.33	10.8
25.0	0.693	1.35	1.96	2.56	3.71	4.84	5.40	6.51	7.61	8.70	10.9	12.5
10.0	1.15	1.95	2.66	3.34	4.64	5.89	6.50	7.70	8.89	10.1	12.4	14.1
5.0	1.50	2.37	3.15	3.88	5.26	6.57	7.22	8.48	9.72	10.9	13.3	15.1
1.0	2.30	3.32	4.20	5.02	6.55	8.00	8.70	10.1	11.4	12.7	15.3	17.2
	0.10	0.40	0.65	1.0	1.5	X	2.5	X	4.0	X	6.5	X
	Acceptable Quality Levels (tightened inspection)											

Note: All values given in above table based on Poisson distribution as an approximation to the Binomial.

TABLE X-L-2 - SAMPLING PLANS FOR SAMPLE SIZE CODE LETTER: L

Type of sampling plan	Cumulative sample size	Acceptable Quality Levels (normal inspection)																																		Cumulative sample size
		Less than 0.065		0.065		0.10		X		0.15		0.25		0.40		0.65		1.0		1.5		X		2.5		X		4.0		X		6.5		Higher than 6.5		
		Ac	Re	Ac	Re	Ac	Re	Ac	Re	Ac	Re	Ac	Re	Ac	Re	Ac	Re	Ac	Re	Ac	Re	Ac	Re	Ac	Re	Ac	Re	Ac	Re	Ac	Re	Ac	Re	Ac	Re	
Single	200	▽		0	1	Use Letter K		Use Letter N		Use Letter M		1	2	2	3	3	4	5	6	7	8	8	9	10	11	12	13	14	15	18	19	21	22	△		200
Double	125	▽		•								0	2	0	3	1	4	2	5	3	7	3	7	5	9	6	10	7	11	9	14	11	16	△		125
	250											1	2	3	4	4	5	6	7	8	9	11	12	12	13	15	16	18	19	23	24	26	27			250
Multiple	50	▽		•								#	2	#	2	#	3	#	4	0	4	0	4	0	5	0	6	1	7	1	8	2	9	△		50
	100											#	2	0	3	0	3	1	5	1	6	2	7	3	8	3	9	4	10	6	12	7	14			100
	150											0	2	0	3	1	4	2	6	3	8	4	9	6	10	7	12	8	13	11	17	13	19			150
	200											0	3	1	4	2	5	3	7	5	10	6	11	8	13	10	15	12	17	16	22	19	25			200
	250											1	3	2	4	3	6	5	8	7	11	9	12	11	15	14	17	17	20	22	25	25	29			250
	300											1	3	3	5	4	6	7	9	10	12	12	14	14	17	18	20	21	23	27	29	31	33			300
	350											2	3	4	5	6	7	9	10	13	14	14	15	18	19	21	22	25	26	32	33	37	38			350
		Less than 0.10		0.10		X		0.15		0.25		0.40		0.65		1.0		1.5		X		2.5		X		4.0		X		6.5		X		Higher than 6.5		
		Acceptable Quality Levels (tightened inspection)																																		

△ = Use next preceding sample size code letter for which acceptance and rejection numbers are available.
▽ = Use next subsequent sample size code letter for which acceptance and rejection numbers are available.
Ac = Acceptance number
Re = Rejection number
• = Use single sampling plan above (or alternatively use letter P).
= Acceptance not permitted at this sample size.

L

M

TABLE X-M—Tables for sample size code letter: M

CHART M - OPERATING CHARACTERISTIC CURVES FOR SINGLE SAMPLING PLANS

(Curves for double and multiple sampling are matched as closely as practicable)

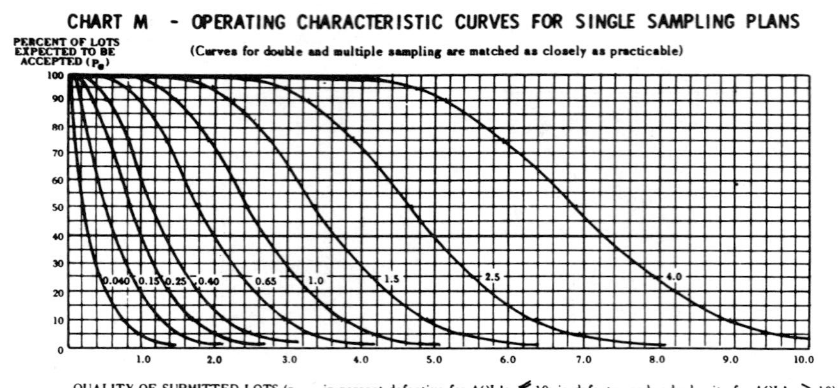

QUALITY OF SUBMITTED LOTS (p, in percent defective for AQL's ≤ 10; in defects per hundred units for AQL's > 10)

Note: Figures on curves are Acceptable Quality Levels (AQL's) for normal inspection.

TABLE X-M-1 - TABULATED VALUES FOR OPERATING CHARACTERSTIC CURVES FOR SINGLE SAMPLING PLANS

P_a	Acceptable Quality Levels (normal inspection)											
	0.040	0.15	0.25	0.40	0.65	1.0	X	1.5	X	2.5	X	4.0
	p (in percent defective or in dejects per hundred units)											
99.0	0.0032	0.047	0.138	0.261	0.566	0.922	1.11	1.51	1.94	2.38	3.28	3.99
95.0	0.0163	0.112	0.259	0.433	0.829	1.26	1.49	1.96	2.44	2.94	3.95	4.73
90.0	0.0333	0.168	0.349	0.533	1.00	1.48	1.72	2.23	2.75	3.27	4.34	5.16
75.0	0.0914	0.305	0.580	0.804	1.34	1.89	2.17	2.74	3.31	3.89	5.05	5.93
50.0	0.220	0.532	0.848	1.17	1.80	2.43	2.75	3.39	4.02	4.66	5.93	6.88
25.0	0.440	0.854	1.24	1.62	2.36	3.07	3.43	4.13	4.83	5.52	6.90	7.92
10.0	0.731	1.23	1.69	2.12	2.94	3.74	4.13	4.89	5.65	6.39	7.86	8.95
5.0	0.951	1.51	2.00	2.46	3.34	4.17	4.58	5.38	6.17	6.95	8.47	9.60
1.0	1.46	2.11	2 67	3.19	4.16	5.08	5.53	6.40	7.25	8.08	9.71	10.9
	0.065	0.25	0.40	0.65	1.0	X	1.5	X	2.5	X	4.0	X
	Acceptable Quality Levels (tightened inspection)											

Note: All values given in above table based on Poisson distribution as an approximation to the Binomial.

TABLE X-M-2 - SAMPLING PLANS FOR SAMPLE SIZE CODE LETTER: M

Type of sampling plan	Cumulative sample size	Acceptable Quality Levels (normal inspection)																																		Cumulative sample size
		Less than 0.040		0.040		0.065		X		0.10		0.15		0.25		0.40		0.65		1.0		X		1.5		X		2.5		X		4.0		Higher than 4.0		
		Ac	Re	Ac	Re	Ac	Re	Ac	Re	Ac	Re	Ac	Re	Ac	Re	Ac	Re	Ac	Re	Ac	Re	Ac	Re	Ac	Re	Ac	Re	Ac	Re	Ac	Re	Ac	Re	Ac	Re	
Single	315	∇		0	1	Use Letter L		Use Letter P		Use Letter N		1	2	2	3	3	4	5	6	7	8	8	9	10	11	12	13	14	15	18	19	21	22	Δ		315
Double	200	∇		*								0	2	0	3	1	4	2	5	3	7	3	7	5	9	6	10	7	11	9	14	11	16	Δ		200
	400											1	2	3	4	4	5	6	7	8	9	11	12	12	13	15	16	18	19	23	24	26	27			400
Multiple	80	∇		*								#	2	#	2	#	3	#	4	0	4	0	4	0	5	0	6	1	7	1	8	2	9	Δ		80
	160											#	2	0	3	0	3	1	5	1	6	2	7	3	8	3	9	4	10	6	12	7	14			160
	240											0	2	0	3	1	4	2	6	3	8	4	9	6	10	7	12	8	13	11	17	13	19			240
	320											0	3	1	4	2	5	3	7	5	10	6	11	8	13	10	15	12	17	16	22	19	25			320
	400											1	3	2	4	3	6	5	8	7	11	9	12	11	15	14	17	17	20	22	25	25	29			400
	480											1	3	3	5	4	6	7	9	10	12	12	14	14	17	18	20	21	23	27	29	31	33			480
	560											2	3	4	5	6	7	9	10	13	14	14	15	18	19	21	22	25	26	32	33	37	38			560
		Less than 0.065		0.065		X		0.10		0.15		0.25		0.40		0.65		1.0		X		1.5		X		2.5		X		4.0		X		Higher than 4.0		
		Acceptable Quality Levels (tightened inspection)																																		

Δ = Use next preceding sample size code letter for which acceptance and rejection numbers are available.
∇ = Use next subsequent sample size code letter for which acceptance and rejection numbers are available.
Ac = Acceptance number.
He = Rejection number.
• = Use single sampling plan above (or alternatively use letter Q).
= Acceptance not permitted at this sample size.

M

N

TABLE X-N—Tables for sample size code letter: N

CHART N - OPERATING CHARACTERISTIC CURVES FOR SINGLE SAMPLING PLANS

(Curves for double and multiple sampling are matched as closely as practicable)

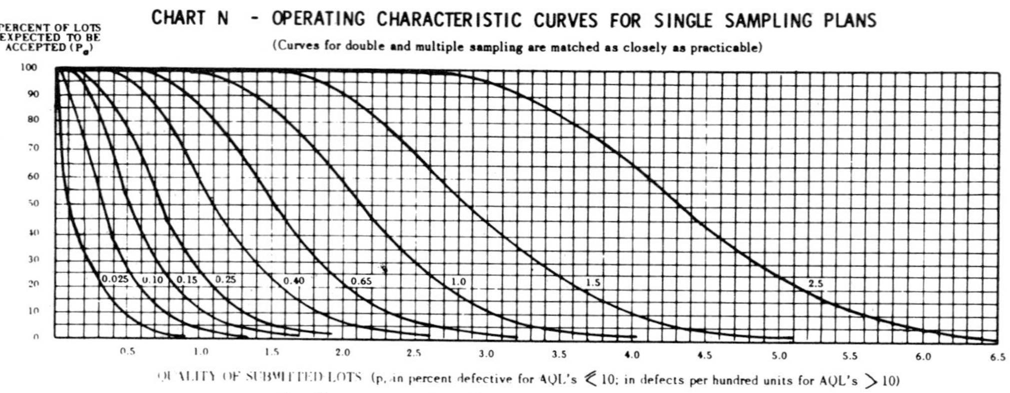

QUALITY OF SUBMITTED LOTS (p, in percent defective for AQL's ≤ 10; in defects per hundred units for AQL's > 10)

Note: Figures on curves are Acceptable Quality Levels (AQL's) for normal inspection.

TABLE X-N-1 - TABULATED VALUES FOR OPERATING CHARACTERISTIC CURVES FOR SINGLE SAMPLING PLANS

P_a	Acceptable Quality Levels (normal inspection)											
	0.025	0.10	0.15	0.25	0.40	0.65	X	1.0	X	1.5	X	2.5
	p (in percent defective or in defects per hundred units)											
99.0	0.0020	0.030	0.087	0.165	0.357	0.581	0.701	0.954	1.22	1.50	2.07	2.51
95.0	0.0103	0.071	0.164	0.273	0.523	0.796	0.939	1.23	1.54	1.85	2.49	2.98
90.0	0.0210	0.106	0.220	0.349	0.630	0.931	1.09	1.40	1.73	2.06	2.73	3.25
75.0	0.0576	0.192	0.345	0.507	0.844	1.19	1.37	1.72	2.08	2.45	3.18	3.74
50.0	0.139	0.336	0.535	0.734	1.13	1.53	1.73	2.13	2.53	2.93	3.73	4.33
25.0	0.277	0.539	0.784	1.02	1.48	1.94	2.16	2.60	3.04	3.48	4.35	4.99
10.0	0.461	0.778	1.06	1.34	1.86	2.35	2.60	3.08	3.56	4.03	4.95	5.64
5.0	0.599	0.949	1.26	1.55	2.10	2.63	2.89	3.39	3.89	4.38	5.34	6.05
1.0	0.921	1.328	1.68	2.01	2.62	3.20	3.48	4.03	4.56	5.09	6.12	6.87
	0.040	0.15	0.25	0.40	0.65	X	1.0	X	1.5	X	2.5	X
	Acceptable Quality Levels (tightened inspection)											

Note: All values given in above table based on Poisson distribution as an approximation to the Binomial

TABLE X-N-2 - SAMPLING PLANS FOR SAMPLE SIZE CODE LETTER: N

Type of sampling plan	Cumulative sample size	Acceptable Quality Levels (normal inspection)																																		Cumulative sample size
		Less than 0.025		0.025		0.040		X		0.065		0.10		0.15		0.25		0.40		0.65		X		1.0		X		1.5		X		2.5		Higher than 2.5		
		Ac	Re	Ac	Re	Ac	Re	Ac	Re	Ac	Re	Ac	Re	Ac	Re	Ac	Re	Ac	Re	Ac	Re	Ac	Re	Ac	Re	Ac	Re	Ac	Re	Ac	Re	Ac	Re	Ac	Re	
Single	500	▽		0	1	Use Letter M		Use Letter Q		Use Letter P		1	2	2	3	3	4	5	6	7	8	8	9	10	11	12	13	14	15	18	19	21	22	△		500
Double	315	▽		•								0	2	0	3	1	4	2	5	3	7	3	7	5	9	6	10	7	11	9	14	11	16	△		315
	630											1	2	3	4	4	5	6	7	8	9	11	12	12	13	15	16	18	19	23	24	26	27			630
Multiple	125	▽		•								#	2	#	2	#	3	#	4	0	4	0	4	0	5	0	6	1	7	1	8	2	9	△		125
	250											#	2	0	3	0	3	1	5	1	6	2	7	3	8	3	9	4	10	6	12	7	14			250
	375											0	2	0	3	1	4	2	6	3	8	4	9	6	10	7	12	8	13	11	17	13	19			375
	500											0	3	1	4	2	5	3	7	5	10	6	11	8	13	10	15	12	17	16	22	19	25			500
	625											1	3	2	4	3	6	5	8	7	11	9	12	11	15	14	17	17	20	22	25	25	29			625
	750											1	3	3	5	4	6	7	9	10	12	12	14	14	17	18	20	21	23	27	29	31	33			750
	875											2	3	4	5	6	7	9	10	13	14	14	15	18	19	21	22	25	26	32	33	37	38			875
		Less than 0.040		0.040		X		0.065		0.10		0.15		0.25		0.40		0.65		X		1.0		X		1.5		X		2.5		X		Higher than 2.5		
		Acceptable Quality Levels (tightened inspection)																																		

△ = Use next preceding sample size code letter for which acceptance and rejection numbers are available.

▽ = Use next subsequent sample size code letter for which acceptance and rejection numbers are available.

Ac = Acceptance number

Re = Rejection number

• = Use single sampling plan above (or alternatively use letter R).

= Acceptance not permitted at this sample size.

N

MIL-STD-414

TABLE A-1

AQL Conversion Table

For specified AQL values falling within these ranges	Use this AQL value
—— to 0.049	0.04
0.050 to 0.069	0.065
0.070 to 0.109	0.10
0.110 to 0.164	0.15
0.165 to 0.279	0.25
0.280 to 0.439	0.40
0.440 to 0.699	0.65
0.700 to 1.09	1.0
1.10 to 1.64	1.5
1.65 to 2.79	2.5
2.80 to 4.39	4.0
4.40 to 6.99	6.5
7.00 to 10.9	10.0
11.00 to 16.4	15.0

TABLE A-2

Sample Size Code Letters[1]

Lot Size	Inspection Levels				
	I	II	III	IV	V
3 to 8	B	B	B	B	C
9 to 15	B	B	B	B	D
16 to 25	B	B	B	C	E
26 to 40	B	B	B	D	F
41 to 65	B	B	C	E	G
66 to 110	B	B	D	F	H
111 to 180	B	C	E	G	I
181 to 300	B	D	F	H	J
301 to 500	C	E	G	I	K
501 to 800	D	F	H	J	L
801 to 1,300	E	G	I	K	L
1,301 to 3,200	F	H	J	L	M
3,201 to 8,000	G	I	L	M	N
8,001 to 22,000	H	J	M	N	O
22,001 to 110,000	I	K	N	O	P
110,001 to 550,000	I	K	O	P	Q
550,001 and over	I	K	P	Q	Q

[1]Sample size code letters given in body of table are applicable when the indicated inspection levels are to be used.

Standard Deviation Method

TABLE B-3

Master Table for Normal and Tightened Inspection for Plans Based on Variability Unknown
(Double Specification Limit and Form 2—Single Specification Limit)

Sample size code letter	Sample size	Acceptable Quality Levels (normal inspection)													
		.04	.065	.10	.15	.25	.40	.65	1.00	1.50	2.50	4.00	6.50	10.00	15.00
		M	M	M	M	M	M	M	M	M	M	M	M	M	M
B	3	↓	↓	↓	↓	↓	↓	↓	▼	▼	7.59	18.86	26.94	33.69	40.47
C	4								1.53	5.50	10.92	16.45	22.86	29.45	36.90
D	5							1.33	3.32	5.83	9.80	14.39	20.19	26.56	33.99
E	7					0.422	1.06	2.14	3.55	5.35	8.40	12.20	17.35	23.29	30.50
F	10				0.349	0.716	1.30	2.17	3.26	4.77	7.29	10.54	15.17	20.74	27.57
G	15	0.099	0.186	0.312	0.503	0.818	1.31	2.11	3.05	4.31	6.56	9.46	13.71	18.94	25.61
H	20	0.135	0.228	0.365	0.544	0.846	1.29	2.05	2.95	4.09	6.17	8.92	12.99	18.03	24.53
I	25	0.155	0.250	0.380	0.551	0.877	1.29	2.00	2.86	3.97	5.97	8.63	12.57	17.51	23.97
J	30	0.179	0.280	0.413	0.581	0.879	1.29	1.98	2.83	3.91	5.86	8.47	12.36	17.24	23.58
K	35	0.170	0.264	0.388	0.535	0.847	1.23	1.87	2.68	3.70	5.57	8.10	11.87	16.65	22.91
L	40	0.179	0.275	0.401	0.566	0.873	1.26	1.88	2.71	3.72	5.58	8.09	11.85	16.61	22.86
M	50	0.163	0.250	0.363	0.503	0.789	1.17	1.71	2.49	3.45	5.20	7.61	11.23	15.87	22.00
N	75	0.147	0.228	0.330	0.467	0.720	1.07	1.60	2.29	3.20	4.87	7.15	10.63	15.13	21.11
O	100	0.145	0.220	0.317	0.447	0.689	1.02	1.53	2.20	3.07	4.69	6.91	10.32	14.75	20.66
P	150	0.134	0.203	0.293	0.413	0.638	0.949	1.43	2.05	2.89	4.43	6.57	9.88	14.20	20.02
Q	200	0.135	0.204	0.294	0.414	0.637	0.945	1.42	2.04	2.87	4.40	6.53	9.81	14.12	19.92
		.065	.10	.15	.25	.40	.65	1.00	1.50	2.50	4.00	6.50	10.00	15.00	
		Acceptability Quality Levels (tightened inspection)													

All AQL and table values are in percent defective.

Use first sampling plan below arrow, that is, both sample size as well as M value. When sample size equals or exceeds lot size, every item in the lot must be inspected.

TABLE B-4

Standard Deviation Method

Master Table for Reduced Inspection for Plans Based on Variability Unknown
(Double Specification Limit and Form 2—Single Specification Limit)

Sample size code letter	Sample size	Acceptable Quality Levels												
		.04	.065	.10	.15	.25	.40	.65	1.00	1.50	2.50	4.00	6.50	10.00
		M	M	M	M	M	M	M	M	M	M	M	M	M
B	3	↓	↓	↓	↓	↓	↓	↓	↓	7.59	18.86	26.94	33.69	40.47
C	3									7.59	18.86	26.94	33.69	40.47
D	3									7.59	18.86	26.94	33.69	40.47
E	3									7.59	18.86	26.94	33.69	40.47
F	4							1.53	5.50	10.92	16.45	22.86	29.45	36.90
G	5						1.33	3.32	5.83	9.80	14.39	20.19	26.56	33.99
H	7				0.422	1.06	2.14	3.55	5.35	8.40	12.20	17.35	23.29	30.50
I	10			0.349	0.716	1.30	2.17	3.26	4.77	7.29	10.54	15.17	20.74	27.57
J	10			0.349	0.716	1.30	2.17	3.26	4.77	7.29	10.54	15.17	20.74	27.57
K	15	0.186	0.312	0.503	0.818	1.31	2.11	3.05	4.31	6.56	9.46	13.71	18.94	25.61
L	20	0.228	0.365	0.544	0.846	1.29	2.05	2.95	4.09	6.17	8.92	12.99	18.03	24.53
M	20	0.228	0.365	0.544	0.846	1.29	2.05	2.95	4.09	6.17	8.92	12.99	18.03	24.53
N	25	0.250	0.380	0.551	0.877	1.29	2.00	2.86	3.97	5.97	8.63	12.57	17.51	23.97
O	30	0.280	0.413	0.581	0.879	1.29	1.98	2.83	3.91	5.86	8.47	12.36	17.24	23.58
P	50	0.250	0.363	0.503	0.789	1.17	1.71	2.49	3.45	5.20	7.61	11.23	15.87	22.00
Q	75	0.228	0.330	0.467	0.720	1.07	1.60	2.29	3.20	4.87	7.15	10.63	15.13	21.11

All AQL and table values are in percent defective.

↓ Use first sampling plan below arrow, that is, both sample size as well as M value. When sample size equals or exceeds lot size, every item in the lot must be inspected.

TABLE B-5

Table for Estimating the Lot Percent Defective Using Standard Deviation Method[1]

Q_U or Q_L	Sample Size 3	4	5	7	10	15	20	25	30	35	40	50	75	100	150	200
0	50.00	50.00	50.00	50.00	50.00	50.00	50.00	50.00	50.00	50.00	50.00	50.00	50.00	50.00	50.00	50.00
.1	47.24	46.67	46.44	46.26	46.16	46.10	46.08	46.06	46.05	46.05	46.04	46.04	46.03	46.03	46.02	46.02
.2	44.46	43.33	42.90	42.54	42.35	42.24	42.19	42.16	42.15	42.13	42.13	42.11	42.10	42.09	42.08	42.08
.3	41.63	40.00	39.37	38.87	38.60	38.44	38.37	38.33	38.31	38.29	38.28	38.27	38.25	38.24	38.22	38.22
.31	41.35	39.67	39.02	38.50	38.23	38.06	37.99	37.95	37.93	37.91	37.90	37.89	37.87	37.86	37.84	37.84
.32	41.06	39.33	38.67	38.14	37.86	37.69	37.62	37.58	37.55	37.54	37.52	37.51	37.49	37.48	37.46	37.46
.33	40.77	39.00	38.32	37.78	37.49	37.31	37.24	37.20	37.18	37.16	37.15	37.13	37.11	37.10	37.09	37.08
.34	40.49	38.67	37.97	37.42	37.12	36.94	36.87	36.83	36.80	36.78	36.77	36.75	36.73	36.72	36.71	36.71
.35	40.20	38.33	37.62	37.06	36.75	36.57	36.49	36.45	36.43	36.41	36.40	36.38	36.36	36.35	36.33	36.33
.36	39.91	38.00	37.28	36.69	36.38	36.20	36.12	36.08	36.05	36.04	36.02	36.01	35.98	35.97	35.96	35.96
.37	39.62	37.67	36.93	36.33	36.02	35.83	35.75	35.71	35.68	35.66	35.65	35.63	35.61	35.60	35.59	35.58
.38	39.33	37.33	36.58	35.98	35.65	35.46	35.38	35.34	35.31	35.29	35.28	35.26	35.24	35.23	35.22	35.21
.39	39.00	37.00	36.23	35.62	35.29	35.10	35.01	34.97	34.94	34.93	34.91	34.89	34.87	34.86	34.85	34.84
.40	38.74	36.67	35.88	35.26	34.93	34.73	34.65	34.60	34.58	34.56	34.54	34.53	34.50	34.49	34.48	34.47
.41	38.45	36.33	35.54	34.90	34.57	34.37	34.28	34.24	34.21	34.19	34.18	34.16	34.13	34.12	34.11	34.10
.42	38.15	36.00	35.19	34.55	34.21	34.00	33.92	33.87	33.85	33.83	33.82	33.79	33.77	33.76	33.74	33.74
.43	37.85	35.67	34.85	34.19	33.85	33.64	33.56	33.51	33.48	33.46	33.45	33.43	33.40	33.39	33.38	33.37
.44	37.56	35.33	34.50	33.84	33.49	33.28	33.20	33.15	33.12	33.10	33.09	33.07	33.04	33.03	33.02	33.01
.45	37.26	35.00	34.16	33.49	33.13	32.92	32.84	32.79	32.76	32.74	32.73	32.71	32.68	32.67	32.66	32.65
.46	36.96	34.67	33.81	33.13	32.78	32.57	32.48	32.43	32.40	32.38	32.37	32.35	32.32	32.31	32.30	32.29
.47	36.66	34.33	33.47	32.78	32.42	32.21	32.12	32.07	32.04	32.02	32.01	31.99	31.96	31.95	31.94	31.93
.48	36.35	34.00	33.12	32.43	32.07	31.85	31.77	31.72	31.69	31.67	31.65	31.63	31.61	31.60	31.58	31.58
.49	36.05	33.67	32.78	32.08	31.72	31.50	31.41	31.36	31.33	31.31	31.30	31.28	31.25	31.24	31.23	31.22
.50	35.75	33.33	32.44	31.74	31.37	31.15	31.06	31.01	30.98	30.96	30.95	30.93	30.90	30.89	30.87	30.87
.51	35.44	33.00	32.10	31.39	31.02	30.80	30.71	30.66	30.63	30.61	30.60	30.57	30.55	30.54	30.52	30.52
.52	35.13	32.67	31.76	31.04	30.67	30.45	30.36	30.31	30.28	30.26	30.25	30.23	30.20	30.19	30.17	30.17
.53	34.82	32.33	31.42	30.70	30.32	30.10	30.01	29.96	29.93	29.91	29.90	29.88	29.85	29.84	29.83	29.82
.54	34.51	32.00	31.08	30.36	29.98	29.76	29.67	29.62	29.59	29.57	29.55	29.53	29.51	29.49	29.48	29.48
.55	34.20	31.67	30.74	30.01	29.64	29.41	29.32	29.27	29.24	29.22	29.21	29.19	29.16	29.15	29.14	29.13
.56	33.88	31.33	30.40	29.67	29.29	29.07	28.98	28.93	28.90	28.88	28.87	28.85	28.82	28.81	28.79	28.79
.57	33.57	31.00	30.06	29.33	28.95	28.73	28.64	28.59	28.56	28.54	28.53	28.51	28.48	28.47	28.45	28.45
.58	33.25	30.67	29.73	28.99	28.61	28.39	28.30	28.25	28.22	28.20	28.19	28.17	28.14	28.13	28.12	28.11
.59	32.93	30.33	29.39	28.66	28.28	28.05	27.96	27.92	27.89	27.87	27.85	27.83	27.81	27.79	27.78	27.77
.60	32.61	30.00	29.05	28.32	27.94	27.72	27.63	27.58	27.55	27.53	27.52	27.50	27.47	27.46	27.45	27.44
.61	32.28	29.67	28.72	27.98	27.60	27.39	27.30	27.25	27.22	27.20	27.18	27.16	27.14	27.13	27.11	27.11
.62	31.96	29.33	28.39	27.65	27.27	27.05	26.96	26.92	26.89	26.87	26.85	26.83	26.81	26.80	26.78	26.78
.63	31.63	29.00	28.05	27.32	26.94	26.72	26.63	26.59	26.56	26.54	26.52	26.50	26.48	26.47	26.45	26.45
.64	31.30	28.67	27.72	26.99	26.61	26.39	26.31	26.26	26.23	26.21	26.20	26.18	26.15	26.14	26.13	26.12
.65	30.97	28.33	27.39	26.66	26.28	26.07	25.98	25.93	25.90	25.88	25.87	25.85	25.83	25.82	25.80	25.80
.66	30.63	28.00	27.06	26.33	25.96	25.74	25.66	25.61	25.58	25.56	25.55	25.53	25.51	25.49	25.48	25.48
.67	30.30	27.67	26.73	26.00	25.63	25.42	25.33	25.29	25.26	25.24	25.23	25.21	25.19	25.17	25.16	25.16
.68	29.96	27.33	26.40	25.68	25.31	25.10	25.01	24.97	24.94	24.92	24.91	24.89	24.87	24.86	24.84	24.84
.69	29.61	27.00	26.07	25.35	24.99	24.78	24.70	24.65	24.62	24.60	24.59	24.57	24.55	24.54	24.53	24.52

[1]Values tabulated are read in percent.

Q_U or Q_L	Sample Size 3	4	5	7	10	15	20	25	30	35	40	50	75	100	150	200
.70	29.27	26.67	25.74	25.03	24.67	24.46	24.38	24.33	24.31	24.29	24.28	24.26	24.24	24.23	24.21	24.21
.71	28.92	26.33	25.41	24.71	24.35	24.15	24.06	24.02	23.99	23.98	23.96	23.95	23.92	23.91	23.90	23.90
.72	28.57	26.00	25.09	24.39	24.03	23.83	23.75	23.71	23.68	23.67	23.65	23.64	23.61	23.60	23.59	23.59
.73	28.22	25.67	24.76	24.07	23.72	23.52	23.44	23.40	23.37	23.36	23.34	23.33	23.31	23.30	23.29	23.28
.74	27.86	25.33	24.44	23.75	23.41	23.21	23.13	23.09	23.07	23.05	23.04	23.02	23.00	22.99	22.98	22.98
.75	27.50	25.00	24.11	23.44	23.10	22.90	22.83	22.79	22.76	22.75	22.73	22.72	22.70	22.69	22.68	22.67
.76	27.13	24.67	23.79	23.12	22.79	22.60	22.52	22.48	22.46	22.44	22.43	22.42	22.40	22.39	22.38	22.37
.77	26.77	24.33	23.47	22.81	22.48	22.30	22.22	22.18	22.16	22.14	22.13	22.12	22.10	22.09	22.08	22.08
.78	26.39	24.00	23.15	22.50	22.18	21.99	21.92	21.89	21.86	21.85	21.84	21.82	21.80	21.79	21.78	21.78
.79	26.02	23.67	22.83	22.19	21.87	21.70	21.63	21.59	21.57	21.55	21.54	21.53	21.51	21.50	21.49	21.49
.80	25.64	23.33	22.51	21.88	21.57	21.40	21.33	21.29	21.27	21.26	21.25	21.23	21.22	21.21	21.20	21.20
.81	25.25	23.00	22.19	21.58	21.27	21.10	21.04	21.00	20.98	20.97	20.96	20.94	20.93	20.92	20.91	20.91
.82	24.86	22.67	21.87	21.27	20.98	20.81	20.75	20.71	20.69	20.68	20.67	20.65	20.64	20.63	20.62	20.62
.83	24.47	22.33	21.56	20.97	20.68	20.52	20.46	20.42	20.40	20.39	20.38	20.37	20.35	20.35	20.34	20.34
.84	24.07	22.00	21.24	20.67	20.39	20.23	20.17	20.14	20.12	20.11	20.10	20.09	20.07	20.06	20.06	20.05
.85	23.67	21.67	20.93	20.37	20.10	19.94	19.89	19.86	19.84	19.82	19.82	19.80	19.79	19.78	19.78	19.77
.86	23.26	21.33	20.62	20.07	19.81	19.66	19.60	19.57	19.56	19.54	19.54	19.53	19.51	19.51	19.50	19.50
.87	22.84	21.00	20.31	19.78	19.52	19.38	19.32	19.30	19.28	19.27	19.26	19.25	19.24	19.23	19.22	19.22
.88	22.42	20.67	20.00	19.48	19.23	19.10	19.04	19.02	19.00	18.99	18.98	18.98	18.96	18.96	18.95	18.95
.89	21.99	20.33	19.69	19.19	18.95	18.82	18.77	18.74	18.73	18.72	18.71	18.70	18.69	18.69	18.68	18.68
.90	21.55	20.00	19.38	18.90	18.67	18.54	18.50	18.47	18.46	18.45	18.44	18.43	18.42	18.42	18.41	18.41
.91	21.11	19.67	19.07	18.61	18.39	18.27	18.22	18.20	18.19	18.18	18.17	18.17	18.16	18.15	18.15	18.15
.92	20.66	19.33	18.77	18.33	18.11	18.00	17.96	17.94	17.92	17.92	17.91	17.90	17.89	17.89	17.88	17.88
.93	20.20	19.00	18.46	18.04	17.84	17.73	17.69	17.67	17.66	17.65	17.65	17.64	17.63	17.63	17.62	17.62
.94	19.74	18.67	18.16	17.76	17.57	17.46	17.43	17.41	17.40	17.39	17.39	17.38	17.37	17.37	17.36	17.36
.95	19.25	18.33	17.86	17.48	17.29	17.20	17.17	17.15	17.14	17.13	17.13	17.12	17.12	17.11	17.11	17.11
.96	18.76	18.00	17.56	17.20	17.03	16.94	16.91	16.89	16.88	16.88	16.87	16.87	16.86	16.86	16.86	16.85
.97	18.25	17.67	17.25	16.92	16.76	16.68	16.65	16.63	16.63	16.62	16.62	16.61	16.61	16.61	16.60	16.60
.98	17.74	17.33	16.96	16.65	16.49	16.42	16.39	16.38	16.37	16.37	16.37	16.36	16.36	16.36	16.36	16.36
.99	17.21	17.00	16.66	16.37	16.23	16.16	16.14	16.13	16.12	16.12	16.12	16.12	16.11	16.11	16.11	16.11
1.00	16.67	16.67	16.36	16.10	15.97	15.91	15.89	15.88	15.88	15.87	15.87	15.87	15.87	15.87	15.87	15.87
1.01	16.11	16.33	16.07	15.83	15.72	15.66	15.64	15.63	15.63	15.63	15.63	15.63	15.62	15.62	15.62	15.62
1.02	15.53	16.00	15.78	15.56	15.46	15.41	15.40	15.39	15.39	15.39	15.39	15.38	15.38	15.38	15.38	15.38
1.03	14.93	15.67	15.48	15.30	15.21	15.17	15.15	15.15	15.15	15.15	15.15	15.15	15.15	15.15	15.15	15.15
1.04	14.31	15.33	15.19	15.03	14.96	14.92	14.91	14.91	14.91	14.91	14.91	14.91	14.91	14.91	14.91	14.91
1.05	13.66	15.00	14.91	14.77	14.71	14.68	14.67	14.67	14.67	14.67	14.68	14.68	14.68	14.68	14.68	14.68
1.06	12.98	14.67	14.62	14.51	14.46	14.44	14.44	14.44	14.44	14.44	14.44	14.45	14.45	14.45	14.45	14.45
1.07	12.27	14.33	14.33	14.26	14.22	14.20	14.20	14.21	14.21	14.21	14.21	14.22	14.22	14.22	14.22	14.23
1.08	11.51	14.00	14.05	14.00	13.97	13.97	13.97	13.98	13.98	13.98	13.99	13.99	13.99	14.00	14.00	14.00
1.09	10.71	13.67	13.76	13.75	13.73	13.74	13.74	13.75	13.75	13.76	13.76	13.77	13.77	13.77	13.78	13.78

TABLE B-5—Continued

Table for Estimating the Lot Percent Defective Using Standard Deviation Method

Q_U or Q_L	Sample Size															
	3	4	5	7	10	15	20	25	30	35	40	50	75	100	150	200
1.10	9.84	13.33	13.48	13.49	13.50	13.51	13.52	13.52	13.53	13.54	13.54	13.54	13.55	13.55	13.56	13.56
1.11	8.89	13.00	13.20	13.25	13.26	13.28	13.29	13.30	13.31	13.31	13.32	13.32	13.33	13.34	13.34	13.34
1.12	7.82	12.67	12.93	13.00	13.03	13.05	13.07	13.08	13.09	13.10	13.10	13.11	13.12	13.12	13.12	13.13
1.13	6.60	12.33	12.65	12.73	12.80	12.83	12.85	12.86	12.87	12.88	12.89	12.89	12.90	12.91	12.91	12.92
1.14	5.08	12.00	12.37	12.51	12.57	12.61	12.63	12.65	12.66	12.67	12.67	12.68	12.69	12.70	12.70	12.70
1.15	0.29	11.67	12.10	12.27	12.34	12.39	12.42	12.44	12.45	12.46	12.46	12.47	12.48	12.49	12.49	12.50
1.16	0.00	11.33	11.83	12.03	12.12	12.18	12.21	12.22	12.24	12.25	12.25	12.26	12.28	12.28	12.29	12.29
1.17	0.00	11.00	11.56	11.79	11.90	11.96	12.00	12.02	12.03	12.04	12.05	12.06	12.07	12.08	12.08	12.09
1.18	0.00	10.67	11.29	11.56	11.68	11.75	11.79	11.81	11.82	11.84	11.84	11.85	11.87	11.88	11.88	11.89
1.19	0.00	10.33	11.02	11.33	11.46	11.54	11.58	11.61	11.62	11.63	11.64	11.65	11.67	11.68	11.69	11.69
1.20	0.00	10.00	10.76	11.10	11.24	11.34	11.38	11.41	11.42	11.43	11.44	11.46	11.47	11.48	11.49	11.49
1.21	0.00	9.67	10.50	10.87	11.03	11.13	11.18	11.21	11.22	11.24	11.25	11.26	11.28	11.29	11.30	11.30
1.22	0.00	9.33	10.23	10.65	10.82	10.93	10.98	11.01	11.03	11.04	11.05	11.07	11.09	11.09	11.10	11.11
1.23	0.00	9.00	9.97	10.42	10.61	10.73	10.78	10.81	10.84	10.85	10.86	10.88	10.90	10.91	10.91	10.92
1.24	0.00	8.67	9.72	10.20	10.41	10.53	10.59	10.62	10.64	10.66	10.67	10.69	10.71	10.72	10.73	10.73
1.25	0.00	8.33	9.46	9.98	10.21	10.34	10.40	10.43	10.46	10.47	10.48	10.50	10.52	10.53	10.54	10.55
1.26	0.00	8.00	9.21	9.77	10.00	10.15	10.21	10.25	10.27	10.29	10.30	10.32	10.34	10.35	10.36	10.37
1.27	0.00	7.67	8.96	9.55	9.81	9.96	10.02	10.06	10.09	10.10	10.12	10.13	10.16	10.17	10.18	10.19
1.28	0.00	7.33	8.71	9.34	9.61	9.77	9.84	9.88	9.90	9.92	9.94	9.95	9.98	9.99	10.00	10.01
1.29	0.00	7.00	8.46	9.13	9.42	9.58	9.65	9.70	9.72	9.74	9.76	9.78	9.80	9.82	9.83	9.83
1.30	0.00	6.67	8.21	8.93	9.22	9.40	9.48	9.52	9.55	9.57	9.58	9.60	9.63	9.64	9.65	9.66
1.31	0.00	6.33	7.97	8.72	9.03	9.22	9.30	9.34	9.37	9.39	9.41	9.43	9.46	9.47	9.48	9.49
1.32	0.00	6.00	7.73	8.52	8.85	9.04	9.12	9.17	9.20	9.22	9.24	9.26	9.29	9.30	9.31	9.32
1.33	0.00	5.67	7.49	8.32	8.66	8.86	8.95	9.00	9.03	9.05	9.07	9.09	9.12	9.13	9.15	9.15
1.34	0.00	5.33	7.25	8.12	8.48	8.69	8.78	8.83	8.86	8.88	8.90	8.92	8.95	8.97	8.98	8.99
1.35	0.00	5.00	7.02	7.92	8.30	8.52	8.61	8.66	8.69	8.72	8.74	8.76	8.79	8.81	8.82	8.83
1.36	0.00	4.67	6.79	7.73	8.12	8.35	8.44	8.50	8.53	8.55	8.57	8.60	8.63	8.65	8.66	8.67
1.37	0.00	4.33	6.56	7.54	7.95	8.18	8.28	8.33	8.37	8.39	8.41	8.44	8.47	8.49	8.50	8.51
1.38	0.00	4.00	6.33	7.35	7.77	8.01	8.12	8.17	8.21	8.24	8.25	8.29	8.31	8.33	8.35	8.35
1.39	0.00	3.67	6.10	7.17	7.60	7.85	7.96	8.01	8.05	8.08	8.10	8.12	8.16	8.18	8.19	8.20
1.40	0.00	3.33	5.88	6.98	7.44	7.69	7.80	7.86	7.90	7.92	7.94	7.97	8.01	8.02	8.04	8.05
1.41	0.00	3.00	5.66	6.80	7.27	7.53	7.64	7.70	7.74	7.77	7.79	7.82	7.86	7.87	7.89	7.90
1.42	0.00	2.67	5.44	6.62	7.10	7.37	7.49	7.55	7.59	7.62	7.64	7.67	7.71	7.73	7.74	7.75
1.43	0.00	2.33	5.23	6.45	6.94	7.22	7.34	7.40	7.44	7.47	7.50	7.52	7.56	7.58	7.60	7.61
1.44	0.00	2.00	5.01	6.27	6.78	7.07	7.19	7.26	7.30	7.33	7.35	7.38	7.42	7.44	7.46	7.47
1.45	0.00	1.67	4.81	6.10	6.63	6.92	7.04	7.11	7.15	7.18	7.21	7.24	7.28	7.30	7.31	7.33
1.46	0.00	1.33	4.60	5.93	6.47	6.77	6.90	6.97	7.01	7.04	7.07	7.10	7.14	7.16	7.18	7.19
1.47	0.00	1.00	4.39	5.77	6.32	6.63	6.75	6.83	6.87	6.90	6.93	6.96	7.00	7.02	7.04	7.05
1.48	0.00	.67	4.19	5.60	6.17	6.48	6.61	6.69	6.73	6.77	6.79	6.82	6.86	6.88	6.90	6.91
1.49	0.00	.33	3.99	5.44	6.02	6.34	6.48	6.55	6.60	6.63	6.65	6.69	6.73	6.75	6.77	6.78

Q_U or Q_L	Sample Size															
	3	4	5	7	10	15	20	25	30	35	40	50	75	100	150	200
1.50	0.00	0.00	3.80	5.28	5.87	6.20	6.34	6.41	6.46	6.50	6.52	6.55	6.60	6.62	6.64	6.65
1.51	0.00	0.00	3.61	5.13	5.73	6.06	6.20	6.28	6.33	6.36	6.39	6.42	6.47	6.49	6.51	6.52
1.52	0.00	0.00	3.42	4.97	5.59	5.93	6.07	6.15	6.20	6.23	6.26	6.29	6.34	6.36	6.38	6.39
1.53	0.00	0.00	3.23	4.82	5.45	5.80	5.94	6.02	6.07	6.11	6.13	6.17	6.21	6.24	6.26	6.27
1.54	0.00	0.00	3.05	4.67	5.31	5.67	5.81	5.89	5.95	5.98	6.01	6.04	6.09	6.11	6.13	6.15
1.55	0.00	0.00	2.87	4.52	5.18	5.54	5.69	5.77	5.82	5.86	5.88	5.92	5.97	5.99	6.01	6.02
1.56	0.00	0.00	2.69	4.38	5.05	5.41	5.56	5.65	5.70	5.74	5.76	5.80	5.85	5.87	5.89	5.90
1.57	0.00	0.00	2.52	4.24	4.92	5.29	5.44	5.53	5.58	5.62	5.64	5.68	5.73	5.75	5.78	5.79
1.58	0.00	0.00	2.35	4.10	4.79	5.16	5.32	5.41	5.46	5.50	5.53	5.56	5.61	5.64	5.66	5.67
1.59	0.00	0.00	2.19	3.96	4.66	5.04	5.20	5.29	5.34	5.38	5.41	5.45	5.50	5.52	5.54	5.56
1.60	0.00	0.00	2.03	3.83	4.54	4.92	5.09	5.17	5.23	5.27	5.30	5.33	5.38	5.41	5.43	5.44
1.61	0.00	0.00	1.87	3.69	4.41	4.81	4.97	5.06	5.12	5.16	5.18	5.22	5.27	5.30	5.32	5.33
1.62	0.00	0.00	1.72	3.57	4.30	4.69	4.86	4.95	5.01	5.04	5.07	5.11	5.16	5.19	5.21	5.23
1.63	0.00	0.00	1.57	3.44	4.18	4.58	4.75	4.84	4.90	4.94	4.97	5.01	5.06	5.08	5.11	5.12
1.64	0.00	0.00	1.42	3.31	4.06	4.47	4.64	4.73	4.79	4.83	4.86	4.90	4.95	4.98	5.00	5.01
1.65	0.00	0.00	1.28	3.19	3.95	4.36	4.53	4.62	4.68	4.72	4.75	4.79	4.85	4.87	4.90	4.91
1.66	0.00	0.00	1.15	3.07	3.84	4.25	4.43	4.52	4.58	4.62	4.65	4.69	4.74	4.77	4.80	4.81
1.67	0.00	0.00	1.02	2.95	3.73	4.15	4.32	4.42	4.48	4.52	4.55	4.59	4.64	4.67	4.70	4.71
1.68	0.00	0.00	0.89	2.84	3.62	4.05	4.22	4.32	4.38	4.42	4.45	4.49	4.55	4.57	4.60	4.61
1.69	0.00	0.00	0.77	2.73	3.52	3.94	4.12	4.22	4.28	4.32	4.35	4.39	4.45	4.47	4.50	4.51
1.70	0.00	0.00	0.66	2.62	3.41	3.84	4.02	4.12	4.18	4.22	4.25	4.30	4.35	4.38	4.41	4.42
1.71	0.00	0.00	0.55	2.51	3.31	3.75	3.93	4.02	4.09	4.13	4.16	4.20	4.26	4.29	4.31	4.32
1.72	0.00	0.00	0.45	2.41	3.21	3.65	3.83	3.93	3.99	4.04	4.07	4.11	4.17	4.19	4.22	4.23
1.73	0.00	0.00	0.36	2.30	3.11	3.56	3.74	3.84	3.90	3.94	3.98	4.02	4.08	4.10	4.13	4.14
1.74	0.00	0.00	0.27	2.20	3.02	3.46	3.65	3.75	3.81	3.85	3.89	3.93	3.99	4.01	4.04	4.05
1.75	0.00	0.00	0.19	2.11	2.93	3.37	3.56	3.66	3.72	3.77	3.80	3.84	3.90	3.93	3.95	3.97
1.76	0.00	0.00	0.12	2.01	2.83	3.28	3.47	3.57	3.63	3.68	3.71	3.76	3.81	3.84	3.87	3.88
1.77	0.00	0.00	0.06	1.92	2.74	3.20	3.38	3.48	3.55	3.59	3.63	3.67	3.73	3.76	3.78	3.80
1.78	0.00	0.00	0.02	1.83	2.66	3.11	3.30	3.40	3.47	3.51	3.54	3.59	3.64	3.67	3.70	3.71
1.79	0.00	0.00	0.00	1.74	2.57	3.03	3.21	3.32	3.38	3.43	3.46	3.51	3.56	3.59	3.63	3.63
1.80	0.00	0.00	0.00	1.65	2.49	2.94	3.13	3.24	3.30	3.35	3.38	3.43	3.48	3.51	3.54	3.55
1.81	0.00	0.00	0.00	1.57	2.40	2.86	3.05	3.16	3.22	3.27	3.30	3.35	3.40	3.43	3.46	3.47
1.82	0.00	0.00	0.00	1.49	2.32	2.79	2.98	3.08	3.15	3.19	3.22	3.27	3.33	3.36	3.38	3.40
1.83	0.00	0.00	0.00	1.41	2.25	2.71	2.90	3.00	3.07	3.11	3.15	3.19	3.25	3.28	3.31	3.32
1.84	0.00	0.00	0.00	1.34	2.17	2.63	2.82	2.93	2.99	3.04	3.07	3.12	3.18	3.21	3.23	3.25
1.85	0.00	0.00	0.00	1.26	2.09	2.56	2.75	2.85	2.92	2.97	3.00	3.05	3.10	3.13	3.16	3.17
1.86	0.00	0.00	0.00	1.19	2.02	2.48	2.68	2.78	2.85	2.89	2.93	2.97	3.03	3.06	3.09	3.10
1.87	0.00	0.00	0.00	1.12	1.95	2.41	2.61	2.71	2.78	2.82	2.86	2.90	2.96	2.99	3.02	3.03
1.88	0.00	0.00	0.00	1.06	1.88	2.34	2.54	2.64	2.71	2.75	2.79	2.83	2.89	2.92	2.95	2.96
1.89	0.00	0.00	0.00	0.99	1.81	2.28	2.47	2.57	2.64	2.68	2.72	2.77	2.83	2.85	2.88	2.90

TABLE B-5—Continued

Table for Estimating the Lot Percent Defective Using Standard Deviation Method

Q_U or Q_L	Sample Size															
	3	4	5	7	10	15	20	25	30	35	40	50	75	100	150	200
1.90	0.00	0.00	0.00	0.93	1.75	2.21	2.40	2.51	2.57	2.62	2.65	2.70	2.76	2.79	2.82	2.83
1.91	0.00	0.00	0.00	0.87	1.68	2.14	2.34	2.44	2.51	2.56	2.59	2.63	2.69	2.72	2.75	2.77
1.92	0.00	0.00	0.00	0.81	1.62	2.08	2.27	2.38	2.45	2.49	2.52	2.57	2.63	2.66	2.69	2.70
1.93	0.00	0.00	0.00	0.76	1.56	2.02	2.21	2.32	2.38	2.43	2.46	2.51	2.57	2.60	2.62	2.64
1.94	0.00	0.00	0.00	0.70	1.50	1.96	2.15	2.25	2.32	2.37	2.40	2.45	2.51	2.54	2.56	2.58
1.95	0.00	0.00	0.00	0.65	1.44	1.90	2.09	2.19	2.26	2.31	2.34	2.39	2.45	2.48	2.50	2.52
1.96	0.00	0.00	0.00	0.60	1.38	1.84	2.03	2.14	2.20	2.25	2.28	2.33	2.39	2.42	2.44	2.46
1.97	0.00	0.00	0.00	0.56	1.33	1.78	1.97	2.08	2.14	2.19	2.22	2.27	2.33	2.36	2.39	2.40
1.98	0.00	0.00	0.00	0.51	1.27	1.73	1.92	2.02	2.09	2.13	2.17	2.21	2.27	2.30	2.33	2.34
1.99	0.00	0.00	0.00	0.47	1.22	1.67	1.86	1.97	2.03	2.08	2.11	2.16	2.22	2.25	2.27	2.29
2.00	0.00	0.00	0.00	0.43	1.17	1.62	1.81	1.91	1.98	2.03	2.06	2.10	2.16	2.19	2.22	2.23
2.01	0.00	0.00	0.00	0.39	1.12	1.57	1.76	1.86	1.93	1.97	2.01	2.05	2.11	2.14	2.17	2.18
2.02	0.00	0.00	0.00	0.36	1.07	1.52	1.71	1.81	1.87	1.92	1.95	2.00	2.06	2.09	2.11	2.13
2.03	0.00	0.00	0.00	0.32	1.03	1.47	1.66	1.76	1.82	1.87	1.90	1.95	2.01	2.04	2.06	2.08
2.04	0.00	0.00	0.00	0.29	0.98	1.42	1.61	1.71	1.77	1.82	1.85	1.90	1.96	1.99	2.01	2.03
2.05	0.00	0.00	0.00	0.26	0.94	1.37	1.56	1.66	1.73	1.77	1.80	1.85	1.91	1.94	1.96	1.98
2.06	0.00	0.00	0.00	0.23	0.90	1.33	1.51	1.61	1.68	1.72	1.76	1.80	1.86	1.89	1.92	1.93
2.07	0.00	0.00	0.00	0.21	0.86	1.28	1.47	1.57	1.63	1.68	1.71	1.76	1.81	1.84	1.87	1.88
2.08	0.00	0.00	0.00	0.18	0.82	1.24	1.42	1.52	1.59	1.63	1.66	1.71	1.77	1.79	1.82	1.84
2.09	0.00	0.00	0.00	0.16	0.78	1.20	1.38	1.48	1.54	1.59	1.62	1.66	1.72	1.75	1.78	1.79
2.10	0.00	0.00	0.00	0.14	0.74	1.16	1.34	1.44	1.50	1.54	1.58	1.62	1.68	1.71	1.73	1.75
2.11	0.00	0.00	0.00	0.12	0.71	1.12	1.30	1.39	1.46	1.50	1.53	1.58	1.63	1.66	1.69	1.70
2.12	0.00	0.00	0.00	0.10	0.67	1.08	1.26	1.35	1.42	1.46	1.49	1.54	1.59	1.62	1.65	1.66
2.13	0.00	0.00	0.00	0.08	0.64	1.04	1.22	1.31	1.38	1.42	1.45	1.50	1.55	1.58	1.61	1.62
2.14	0.00	0.00	0.00	0.07	0.61	1.00	1.18	1.28	1.34	1.38	1.41	1.46	1.51	1.54	1.57	1.58
2.15	0.00	0.00	0.00	0.06	0.58	0.97	1.14	1.24	1.30	1.34	1.37	1.42	1.47	1.50	1.53	1.54
2.16	0.00	0.00	0.00	0.05	0.55	0.93	1.10	1.20	1.26	1.30	1.34	1.38	1.43	1.46	1.49	1.50
2.17	0.00	0.00	0.00	0.04	0.52	0.90	1.07	1.16	1.22	1.27	1.30	1.34	1.40	1.42	1.45	1.46
2.18	0.00	0.00	0.00	0.03	0.49	0.87	1.03	1.13	1.19	1.23	1.26	1.30	1.36	1.39	1.41	1.42
2.19	0.00	0.00	0.00	0.02	0.46	0.83	1.00	1.09	1.15	1.20	1.23	1.27	1.32	1.35	1.38	1.39
2.20	0.000	0.000	0.000	0.015	0.437	0.803	0.968	1.061	1.120	1.161	1.192	1.233	1.287	1.314	1.340	1.352
2.21	0.000	0.000	0.000	0.010	0.413	0.772	0.936	1.028	1.087	1.128	1.158	1.199	1.253	1.279	1.305	1.318
2.22	0.000	0.000	0.000	0.006	0.389	0.743	0.905	0.996	1.054	1.095	1.125	1.166	1.219	1.245	1.271	1.283
2.23	0.000	0.000	0.000	0.003	0.366	0.715	0.875	0.965	1.023	1.063	1.093	1.134	1.186	1.212	1.238	1.250
2.24	0.000	0.000	0.000	0.002	0.345	0.687	0.845	0.935	0.992	1.032	1.061	1.102	1.154	1.180	1.205	1.218
2.25	0.000	0.000	0.000	0.001	0.324	0.660	0.816	0.905	0.962	1.002	1.031	1.071	1.123	1.148	1.173	1.186
2.26	0.000	0.000	0.000	0.000	0.304	0.634	0.789	0.876	0.933	0.972	1.001	1.041	1.092	1.117	1.142	1.155
2.27	0.000	0.000	0.000	0.000	0.285	0.609	0.762	0.848	0.904	0.943	0.972	1.011	1.062	1.087	1.112	1.124
2.28	0.000	0.000	0.000	0.000	0.267	0.585	0.735	0.821	0.876	0.915	0.943	0.982	1.033	1.058	1.082	1.094
2.29	0.000	0.000	0.000	0.000	0.250	0.561	0.710	0.794	0.849	0.887	0.915	0.954	1.004	1.029	1.053	1.065

Q_U or Q_L	Sample Size															
	3	4	5	7	10	15	20	25	30	35	40	50	75	100	150	200
2.30	0.000	0.000	0.000	0.000	0.233	0.538	0.685	0.769	0.823	0.861	0.888	0.927	0.977	1.001	1.025	1.037
2.31	0.000	0.000	0.000	0.000	0.218	0.516	0.661	0.743	0.797	0.834	0.862	0.900	0.949	0.974	0.997	1.009
2.32	0.000	0.000	0.000	0.000	0.203	0.495	0.637	0.719	0.772	0.809	0.836	0.874	0.923	0.947	0.971	0.982
2.33	0.000	0.000	0.000	0.000	0.189	0.474	0.614	0.695	0.748	0.784	0.811	0.848	0.897	0.921	0.944	0.956
2.34	0.000	0.000	0.000	0.000	0.175	0.454	0.592	0.672	0.724	0.760	0.787	0.824	0.872	0.895	0.915	0.930
2.35	0.000	0.000	0.000	0.000	0.163	0.435	0.571	0.650	0.701	0.736	0.763	0.799	0.847	0.870	0.893	0.905
2.36	0.000	0.000	0.000	0.000	0.151	0.416	0.550	0.628	0.678	0.714	0.740	0.776	0.823	0.846	0.869	0.880
2.37	0.000	0.000	0.000	0.000	0.139	0.398	0.530	0.606	0.656	0.691	0.717	0.753	0.799	0.822	0.845	0.856
2.38	0.000	0.000	0.000	0.000	0.128	0.381	0.510	0.586	0.635	0.670	0.695	0.730	0.777	0.799	0.822	0.833
2.39	0.000	0.000	0.000	0.000	0.118	0.364	0.491	0.566	0.614	0.648	0.674	0.709	0.754	0.777	0.799	0.810
2.40	0.000	0.000	0.000	0.000	0.109	0.348	0.473	0.546	0.594	0.628	0.653	0.687	0.732	0.755	0.777	0.787
2.41	0.000	0.000	0.000	0.000	0.100	0.332	0.455	0.527	0.575	0.608	0.633	0.667	0.711	0.733	0.755	0.766
2.42	0.000	0.000	0.000	0.000	0.091	0.317	0.437	0.509	0.555	0.588	0.613	0.646	0.691	0.712	0.734	0.744
2.43	0.000	0.000	0.000	0.000	0.083	0.302	0.421	0.491	0.537	0.569	0.593	0.627	0.670	0.692	0.713	0.724
2.44	0.000	0.000	0.000	0.000	0.076	0.288	0.404	0.474	0.519	0.551	0.575	0.608	0.651	0.672	0.693	0.703
2.45	0.000	0.000	0.000	0.000	0.069	0.275	0.389	0.457	0.501	0.533	0.556	0.589	0.632	0.653	0.673	0.684
2.46	0.000	0.000	0.000	0.000	0.063	0.262	0.373	0.440	0.484	0.516	0.539	0.571	0.613	0.634	0.654	0.664
2.47	0.000	0.000	0.000	0.000	0.057	0.249	0.359	0.425	0.468	0.499	0.521	0.553	0.595	0.615	0.635	0.646
2.48	0.000	0.000	0.000	0.000	0.051	0.237	0.344	0.409	0.452	0.482	0.505	0.536	0.577	0.597	0.617	0.627
2.49	0.000	0.000	0.000	0.000	0.046	0.226	0.331	0.394	0.436	0.466	0.488	0.519	0.560	0.580	0.600	0.609
2.50	0.000	0.000	0.000	0.000	0.041	0.214	0.317	0.380	0.421	0.451	0.473	0.503	0.543	0.563	0.582	0.592
2.51	0.000	0.000	0.000	0.000	0.037	0.204	0.304	0.366	0.407	0.436	0.457	0.487	0.527	0.546	0.565	0.575
2.52	0.000	0.000	0.000	0.000	0.033	0.193	0.292	0.352	0.392	0.421	0.442	0.472	0.511	0.530	0.549	0.558
2.53	0.000	0.000	0.000	0.000	0.029	0.184	0.280	0.339	0.379	0.407	0.428	0.457	0.495	0.514	0.533	0.542
2.54	0.000	0.000	0.000	0.000	0.026	0.174	0.268	0.326	0.365	0.393	0.413	0.442	0.480	0.499	0.517	0.527
2.55	0.000	0.000	0.000	0.000	0.023	0.165	0.257	0.314	0.352	0.379	0.400	0.428	0.465	0.484	0.502	0.511
2.56	0.000	0.000	0.000	0.000	0.020	0.156	0.246	0.302	0.340	0.366	0.386	0.414	0.451	0.469	0.487	0.496
2.57	0.000	0.000	0.000	0.000	0.017	0.148	0.236	0.291	0.327	0.354	0.373	0.401	0.437	0.455	0.473	0.482
2.58	0.000	0.000	0.000	0.000	0.015	0.140	0.226	0.279	0.316	0.341	0.361	0.388	0.424	0.441	0.459	0.468
2.59	0.000	0.000	0.000	0.000	0.013	0.133	0.216	0.269	0.304	0.330	0.349	0.375	0.410	0.428	0.445	0.454
2.60	0.000	0.000	0.000	0.000	0.011	0.125	0.207	0.258	0.293	0.318	0.337	0.363	0.398	0.415	0.432	0.441
2.61	0.000	0.000	0.000	0.000	0.009	0.118	0.198	0.248	0.282	0.307	0.325	0.351	0.385	0.402	0.419	0.428
2.62	0.000	0.000	0.000	0.000	0.008	0.112	0.189	0.238	0.272	0.296	0.314	0.339	0.373	0.390	0.406	0.415
2.63	0.000	0.000	0.000	0.000	0.007	0.105	0.181	0.229	0.262	0.285	0.303	0.328	0.361	0.378	0.394	0.402
2.64	0.000	0.000	0.000	0.000	0.005	0.099	0.172	0.220	0.252	0.275	0.293	0.317	0.350	0.366	0.382	0.390
2.65	0.000	0.000	0.000	0.000	0.005	0.094	0.165	0.211	0.243	0.265	0.282	0.307	0.339	0.355	0.371	0.379
2.66	0.000	0.000	0.000	0.000	0.004	0.088	0.157	0.202	0.233	0.256	0.273	0.296	0.328	0.344	0.359	0.367
2.67	0.000	0.000	0.000	0.000	0.003	0.083	0.150	0.194	0.224	0.246	0.263	0.286	0.317	0.333	0.348	0.356
2.68	0.000	0.000	0.000	0.000	0.002	0.078	0.143	0.186	0.216	0.237	0.254	0.277	0.307	0.322	0.338	0.345
2.69	0.000	0.000	0.000	0.000	0.002	0.073	0.136	0.179	0.208	0.229	0.245	0.267	0.297	0.312	0.327	0.335

TABLE B-5—Continued

Table for Estimating the Lot Percent Defective Using Standard Deviation Method

Q_U or Q_L	Sample Size															
	3	4	5	7	10	15	20	25	30	35	40	50	75	100	150	200
2.70	0.000	0.000	0.000	0.000	0.001	0.069	0.130	0.171	0.200	0.220	0.236	0.258	0.288	0.302	0.317	0.325
2.71	0.000	0.000	0.000	0.000	0.001	0.064	0.124	0.164	0.192	0.212	0.227	0.249	0.278	0.293	0.307	0.315
2.72	0.000	0.000	0.000	0.000	0.000	0.060	0.118	0.157	0.184	0.204	0.219	0.241	0.269	0.283	0.298	0.305
2.73	0.000	0.000	0.000	0.000	0.000	0.057	0.112	0.151	0.177	0.197	0.211	0.232	0.260	0.274	0.288	0.296
2.74	0.000	0.000	0.000	0.000	0.000	0.053	0.107	0.144	0.170	0.189	0.204	0.224	0.252	0.266	0.279	0.286
2.75	0.000	0.000	0.000	0.000	0.000	0.049	0.102	0.138	0.163	0.182	0.196	0.216	0.243	0.257	0.271	0.277
2.76	0.000	0.000	0.000	0.000	0.000	0.046	0.097	0.132	0.157	0.175	0.189	0.209	0.235	0.249	0.262	0.269
2.77	0.000	0.000	0.000	0.000	0.000	0.043	0.092	0.126	0.151	0.168	0.182	0.201	0.227	0.241	0.254	0.260
2.78	0.000	0.000	0.000	0.000	0.000	0.040	0.087	0.121	0.145	0.162	0.175	0.194	0.220	0.233	0.246	0.252
2.79	0.000	0.000	0.000	0.000	0.000	0.037	0.083	0.115	0.139	0.156	0.169	0.187	0.212	0.225	0.238	0.244
2.80	0.000	0.000	0.000	0.000	0.000	0.035	0.079	0.110	0.133	0.150	0.162	0.181	0.205	0.218	0.230	0.237
2.81	0.000	0.000	0.000	0.000	0.000	0.032	0.075	0.105	0.128	0.144	0.156	0.174	0.198	0.211	0.223	0.229
2.82	0.000	0.000	0.000	0.000	0.000	0.030	0.071	0.101	0.122	0.138	0.150	0.168	0.192	0.204	0.216	0.222
2.83	0.000	0.000	0.000	0.000	0.000	0.028	0.067	0.096	0.117	0.133	0.145	0.162	0.185	0.197	0.209	0.215
2.84	0.000	0.000	0.000	0.000	0.000	0.026	0.064	0.092	0.112	0.128	0.139	0.156	0.179	0.190	0.202	0.208
2.85	0.000	0.000	0.000	0.000	0.000	0.024	0.060	0.088	0.108	0.122	0.134	0.150	0.173	0.184	0.195	0.201
2.86	0.000	0.000	0.000	0.000	0.000	0.022	0.057	0.084	0.103	0.118	0.129	0.145	0.167	0.178	0.189	0.195
2.87	0.000	0.000	0.000	0.000	0.000	0.020	0.054	0.080	0.099	0.113	0.124	0.139	0.161	0.172	0.183	0.188
2.88	0.000	0.000	0.000	0.000	0.000	0.019	0.051	0.076	0.094	0.108	0.119	0.134	0.155	0.166	0.177	0.182
2.89	0.000	0.000	0.000	0.000	0.000	0.017	0.048	0.073	0.090	0.104	0.114	0.129	0.150	0.160	0.171	0.176
2.90	0.000	0.000	0.000	0.000	0.000	0.016	0.046	0.069	0.087	0.100	0.110	0.125	0.145	0.155	0.165	0.171
2.91	0.000	0.000	0.000	0.000	0.000	0.015	0.043	0.066	0.083	0.096	0.106	0.120	0.140	0.150	0.160	0.165
2.92	0.000	0.000	0.000	0.000	0.000	0.013	0.041	0.063	0.079	0.092	0.101	0.115	0.135	0.145	0.155	0.160
2.93	0.000	0.000	0.000	0.000	0.000	0.012	0.038	0.060	0.076	0.088	0.097	0.111	0.130	0.140	0.149	0.154
2.94	0.000	0.000	0.000	0.000	0.000	0.011	0.036	0.057	0.072	0.084	0.093	0.107	0.125	0.135	0.144	0.149
2.95	0.000	0.000	0.000	0.000	0.000	0.010	0.034	0.054	0.069	0.081	0.090	0.103	0.121	0.130	0.140	0.144
2.96	0.000	0.000	0.000	0.000	0.000	0.009	0.032	0.051	0.066	0.077	0.086	0.099	0.117	0.126	0.135	0.140
2.97	0.000	0.000	0.000	0.000	0.000	0.009	0.030	0.049	0.063	0.074	0.083	0.095	0.112	0.121	0.130	0.135
2.98	0.000	0.000	0.000	0.000	0.000	0.008	0.028	0.046	0.060	0.071	0.079	0.091	0.108	0.117	0.126	0.130
2.99	0.000	0.000	0.000	0.000	0.000	0.007	0.027	0.044	0.057	0.068	0.076	0.088	0.104	0.113	0.122	0.126
3.00	0.000	0.000	0.000	0.000	0.000	0.006	0.025	0.042	0.055	0.065	0.073	0.084	0.101	0.109	0.118	0.122
3.01	0.000	0.000	0.000	0.000	0.000	0.006	0.024	0.040	0.052	0.062	0.070	0.081	0.097	0.105	0.114	0.118
3.02	0.000	0.000	0.000	0.000	0.000	0.005	0.022	0.038	0.050	0.059	0.067	0.078	0.093	0.101	0.110	0.114
3.03	0.000	0.000	0.000	0.000	0.000	0.005	0.021	0.036	0.048	0.057	0.064	0.075	0.090	0.098	0.106	0.110
3.04	0.000	0.000	0.000	0.000	0.000	0.004	0.019	0.034	0.045	0.054	0.061	0.072	0.087	0.094	0.102	0.106
3.05	0.000	0.000	0.000	0.000	0.000	0.004	0.018	0.032	0.043	0.052	0.059	0.069	0.083	0.091	0.099	0.103
3.06	0.000	0.000	0.000	0.000	0.000	0.003	0.017	0.030	0.041	0.050	0.056	0.066	0.080	0.088	0.095	0.099
3.07	0.000	0.000	0.000	0.000	0.000	0.003	0.016	0.029	0.039	0.047	0.054	0.064	0.077	0.085	0.092	0.096
3.08	0.000	0.000	0.000	0.000	0.000	0.003	0.015	0.027	0.037	0.045	0.052	0.061	0.074	0.081	0.089	0.092
3.09	0.000	0.000	0.000	0.000	0.000	0.002	0.014	0.026	0.036	0.043	0.049	0.059	0.072	0.079	0.086	0.089

Q_U or Q_L	Sample Size															
	3	4	5	7	10	15	20	25	30	35	40	50	75	100	150	200
3.10	0.000	0.000	0.000	0.000	0.000	0.002	0.013	0.024	0.034	0.041	0.047	0.056	0.069	0.076	0.083	0.086
3.11	0.000	0.000	0.000	0.000	0.000	0.002	0.012	0.023	0.032	0.039	0.045	0.054	0.066	0.073	0.080	0.083
3.12	0.000	0.000	0.000	0.000	0.000	0.002	0.011	0.022	0.031	0.038	0.043	0.052	0.064	0.070	0.077	0.080
3.13	0.000	0.000	0.000	0.000	0.000	0.002	0.011	0.021	0.029	0.036	0.041	0.050	0.061	0.068	0.074	0.077
3.14	0.000	0.000	0.000	0.000	0.000	0.001	0.010	0.019	0.028	0.034	0.040	0.048	0.059	0.065	0.071	0.075
3.15	0.000	0.000	0.000	0.000	0.000	0.001	0.009	0.018	0.026	0.033	0.038	0.046	0.057	0.063	0.069	0.072
3.16	0.000	0.000	0.000	0.000	0.000	0.001	0.009	0.017	0.025	0.031	0.036	0.044	0.055	0.060	0.066	0.069
3.17	0.000	0.000	0.000	0.000	0.000	0.001	0.008	0.016	0.024	0.030	0.035	0.042	0.053	0.058	0.064	0.067
3.18	0.000	0.000	0.000	0.000	0.000	0.001	0.007	0.015	0.022	0.028	0.033	0.040	0.050	0.056	0.062	0.065
3.19	0.000	0.000	0.000	0.000	0.000	0.001	0.007	0.015	0.021	0.027	0.032	0.038	0.049	0.054	0.059	0.062
3.20	0.000	0.000	0.000	0.000	0.000	0.001	0.006	0.014	0.020	0.026	0.030	0.037	0.047	0.052	0.057	0.060
3.21	0.000	0.000	0.000	0.000	0.000	0.000	0.006	0.013	0.019	0.024	0.029	0.035	0.045	0.050	0.055	0.058
3.22	0.000	0.000	0.000	0.000	0.000	0.000	0.005	0.012	0.018	0.023	0.027	0.034	0.043	0.048	0.053	0.056
3.23	0.000	0.000	0.000	0.000	0.000	0.000	0.005	0.011	0.017	0.022	0.026	0.032	0.041	0.046	0.051	0.054
3.24	0.000	0.000	0.000	0.000	0.000	0.000	0.005	0.011	0.016	0.021	0.025	0.031	0.040	0.044	0.049	0.052
3.25	0.000	0.000	0.000	0.000	0.000	0.000	0.004	0.010	0.015	0.020	0.024	0.030	0.038	0.043	0.048	0.050
3.26	0.000	0.000	0.000	0.000	0.000	0.000	0.004	0.009	0.015	0.019	0.023	0.028	0.037	0.041	0.046	0.048
3.27	0.000	0.000	0.000	0.000	0.000	0.000	0.004	0.009	0.014	0.019	0.022	0.027	0.035	0.040	0.044	0.046
3.28	0.000	0.000	0.000	0.000	0.000	0.000	0.003	0.008	0.013	0.017	0.021	0.026	0.034	0.038	0.042	0.045
3.29	0.000	0.000	0.000	0.000	0.000	0.000	0.003	0.008	0.012	0.016	0.020	0.025	0.032	0.037	0.041	0.043
3.30	0.000	0.000	0.000	0.000	0.000	0.000	0.003	0.007	0.012	0.015	0.019	0.024	0.031	0.035	0.039	0.042
3.31	0.000	0.000	0.000	0.000	0.000	0.000	0.003	0.007	0.011	0.015	0.018	0.023	0.030	0.034	0.038	0.040
3.32	0.000	0.000	0.000	0.000	0.000	0.000	0.002	0.006	0.010	0.014	0.017	0.022	0.029	0.032	0.036	0.039
3.33	0.000	0.000	0.000	0.000	0.000	0.000	0.002	0.006	0.010	0.013	0.016	0.021	0.027	0.031	0.035	0.037
3.34	0.000	0.000	0.000	0.000	0.000	0.000	0.002	0.006	0.009	0.013	0.015	0.020	0.026	0.030	0.034	0.036
3.35	0.000	0.000	0.000	0.000	0.000	0.000	0.002	0.005	0.009	0.012	0.015	0.019	0.025	0.029	0.032	0.034
3.36	0.000	0.000	0.000	0.000	0.000	0.000	0.002	0.005	0.008	0.011	0.014	0.018	0.024	0.028	0.031	0.033
3.37	0.000	0.000	0.000	0.000	0.000	0.000	0.002	0.005	0.008	0.011	0.013	0.017	0.023	0.026	0.030	0.032
3.38	0.000	0.000	0.000	0.000	0.000	0.000	0.001	0.004	0.007	0.010	0.013	0.016	0.022	0.025	0.029	0.031
3.39	0.000	0.000	0.000	0.000	0.000	0.000	0.001	0.004	0.007	0.010	0.012	0.016	0.021	0.024	0.028	0.029
3.40	0.000	0.000	0.000	0.000	0.000	0.000	0.001	0.004	0.007	0.009	0.011	0.015	0.020	0.023	0.027	0.028
3.41	0.000	0.000	0.000	0.000	0.000	0.000	0.001	0.003	0.006	0.009	0.011	0.014	0.020	0.022	0.026	0.027
3.42	0.000	0.000	0.000	0.000	0.000	0.000	0.001	0.003	0.006	0.008	0.010	0.014	0.019	0.022	0.025	0.026
3.43	0.000	0.000	0.000	0.000	0.000	0.000	0.001	0.003	0.005	0.008	0.010	0.013	0.018	0.021	0.024	0.025
3.44	0.000	0.000	0.000	0.000	0.000	0.000	0.001	0.003	0.005	0.007	0.009	0.012	0.017	0.020	0.023	0.024
3.45	0.000	0.000	0.000	0.000	0.000	0.000	0.001	0.003	0.005	0.007	0.009	0.012	0.016	0.019	0.022	0.023
3.46	0.000	0.000	0.000	0.000	0.000	0.000	0.001	0.002	0.005	0.007	0.008	0.011	0.016	0.018	0.021	0.022
3.47	0.000	0.000	0.000	0.000	0.000	0.000	0.001	0.002	0.004	0.006	0.008	0.011	0.015	0.017	0.020	0.022
3.48	0.000	0.000	0.000	0.000	0.000	0.000	0.001	0.002	0.004	0.006	0.007	0.010	0.014	0.017	0.019	0.021
3.49	0.000	0.000	0.000	0.000	0.000	0.000	0.000	0.002	0.004	0.005	0.007	0.010	0.014	0.016	0.019	0.020

TABLE B-5—Continued

Table for Estimating the Lot Percent Defective Using Standard Deviation Method

Q_U or Q_L	Sample Size															
	3	4	5	7	10	15	20	25	30	35	40	50	75	100	150	200
3.50	0.000	0.000	0.000	0.000	0.000	0.000	0.000	0.002	0.003	0.005	0.007	0.009	0.013	0.015	0.018	0.019
3.51	0.000	0.000	0.000	0.000	0.000	0.000	0.000	0.002	0.003	0.005	0.006	0.009	0.013	0.015	0.017	0.018
3.52	0.000	0.000	0.000	0.000	0.000	0.000	0.000	0.002	0.003	0.005	0.006	0.008	0.012	0.014	0.017	0.018
3.53	0.000	0.000	0.000	0.000	0.000	0.000	0.000	0.001	0.003	0.004	0.006	0.008	0.012	0.014	0.016	0.017
3.54	0.000	0.000	0.000	0.000	0.000	0.000	0.000	0.001	0.003	0.004	0.005	0.008	0.011	0.013	0.015	0.016
3.55	0.000	0.000	0.000	0.000	0.000	0.000	0.000	0.001	0.003	0.004	0.005	0.007	0.011	0.012	0.015	0.016
3.56	0.000	0.000	0.000	0.000	0.000	0.000	0.000	0.001	0.002	0.004	0.005	0.007	0.010	0.012	0.014	0.015
3.57	0.000	0.000	0.000	0.000	0.000	0.000	0.000	0.001	0.002	0.003	0.005	0.006	0.010	0.011	0.013	0.014
3.58	0.000	0.000	0.000	0.000	0.000	0.000	0.000	0.001	0.002	0.003	0.004	0.006	0.009	0.011	0.013	0.014
3.59	0.000	0.000	0.000	0.000	0.000	0.000	0.000	0.001	0.002	0.003	0.004	0.006	0.009	0.010	0.012	0.013
3.60	0.000	0.000	0.000	0.000	0.000	0.000	0.000	0.001	0.002	0.003	0.004	0.006	0.008	0.010	0.012	0.013
3.61	0.000	0.000	0.000	0.000	0.000	0.000	0.000	0.001	0.002	0.003	0.004	0.005	0.008	0.010	0.011	0.012
3.62	0.000	0.000	0.000	0.000	0.000	0.000	0.000	0.001	0.002	0.003	0.003	0.005	0.008	0.009	0.011	0.012
3.63	0.000	0.000	0.000	0.000	0.000	0.000	0.000	0.001	0.001	0.002	0.003	0.005	0.007	0.009	0.010	0.011
3.64	0.000	0.000	0.000	0.000	0.000	0.000	0.000	0.001	0.001	0.002	0.003	0.004	0.007	0.008	0.010	0.011
3.65	0.000	0.000	0.000	0.000	0.000	0.000	0.000	0.001	0.001	0.002	0.003	0.004	0.007	0.008	0.010	0.010
3.66	0.000	0.000	0.000	0.000	0.000	0.000	0.000	0.000	0.001	0.002	0.003	0.004	0.006	0.008	0.009	0.010
3.67	0.000	0.000	0.000	0.000	0.000	0.000	0.000	0.000	0.001	0.002	0.003	0.004	0.006	0.007	0.009	0.010
3.68	0.000	0.000	0.000	0.000	0.000	0.000	0.000	0.000	0.001	0.002	0.002	0.004	0.006	0.007	0.008	0.009
3.69	0.000	0.000	0.000	0.000	0.000	0.000	0.000	0.000	0.001	0.002	0.002	0.003	0.005	0.007	0.008	0.009

Q_U or Q_L	Sample Size															
	3	4	5	7	10	15	20	25	30	35	40	50	75	100	150	200
3.70	0.000	0.000	0.000	0.000	0.000	0.000	0.000	0.000	0.001	0.002	0.002	0.003	0.005	0.006	0.008	0.008
3.71	0.000	0.000	0.000	0.000	0.000	0.000	0.000	0.000	0.001	0.001	0.002	0.003	0.005	0.006	0.007	0.008
3.72	0.000	0.000	0.000	0.000	0.000	0.000	0.000	0.000	0.001	0.001	0.002	0.003	0.005	0.006	0.007	0.008
3.73	0.000	0.000	0.000	0.000	0.000	0.000	0.000	0.000	0.001	0.001	0.002	0.003	0.005	0.006	0.007	0.007
3.74	0.000	0.000	0.000	0.000	0.000	0.000	0.000	0.000	0.001	0.001	0.002	0.003	0.004	0.005	0.007	0.007
3.75	0.000	0.000	0.000	0.000	0.000	0.000	0.000	0.000	0.001	0.001	0.002	0.002	0.004	0.005	0.006	0.007
3.76	0.000	0.000	0.000	0.000	0.000	0.000	0.000	0.000	0.001	0.001	0.001	0.002	0.004	0.005	0.006	0.007
3.77	0.000	0.000	0.000	0.000	0.000	0.000	0.000	0.000	0.001	0.001	0.001	0.002	0.004	0.005	0.006	0.006
3.78	0.000	0.000	0.000	0.000	0.000	0.000	0.000	0.000	0.000	0.001	0.001	0.002	0.004	0.004	0.005	0.006
3.79	0.000	0.000	0.000	0.000	0.000	0.000	0.000	0.000	0.000	0.001	0.001	0.002	0.003	0.004	0.005	0.006
3.80	0.000	0.000	0.000	0.000	0.000	0.000	0.000	0.000	0.000	0.001	0.001	0.002	0.003	0.004	0.005	0.006
3.81	0.000	0.000	0.000	0.000	0.000	0.000	0.000	0.000	0.000	0.001	0.001	0.002	0.003	0.004	0.005	0.005
3.82	0.000	0.000	0.000	0.000	0.000	0.000	0.000	0.000	0.000	0.001	0.001	0.002	0.003	0.004	0.005	0.005
3.83	0.000	0.000	0.000	0.000	0.000	0.000	0.000	0.000	0.000	0.001	0.001	0.002	0.003	0.004	0.004	0.005
3.84	0.000	0.000	0.000	0.000	0.000	0.000	0.000	0.000	0.000	0.001	0.001	0.001	0.003	0.003	0.004	0.005
3.85	0.000	0.000	0.000	0.000	0.000	0.000	0.000	0.000	0.000	0.001	0.001	0.001	0.002	0.003	0.004	0.004
3.86	0.000	0.000	0.000	0.000	0.000	0.000	0.000	0.000	0.000	0.000	0.001	0.001	0.002	0.003	0.004	0.004
3.87	0.000	0.000	0.000	0.000	0.000	0.000	0.000	0.000	0.000	0.000	0.001	0.001	0.002	0.003	0.004	0.004
3.88	0.000	0.000	0.000	0.000	0.000	0.000	0.000	0.000	0.000	0.000	0.001	0.001	0.002	0.003	0.004	0.004
3.89	0.000	0.000	0.000	0.000	0.000	0.000	0.000	0.000	0.000	0.000	0.001	0.001	0.002	0.003	0.003	0.004
3.90	0.000	0.000	0.000	0.000	0.000	0.000	0.000	0.000	0.000	0.000	0.001	0.001	0.002	0.003	0.003	0.004

Appendix D

Selected Solutions

NOTE: These solutions are computer generated and may vary slightly from table and calculator-generated solutions.

Chapter One

Problem 1-1

VALUE X	*f*	*BAR*
1.02	0	
1.03	0	
1.04	0	
1.05	3	III
1.06	3	III
1.07	11	IIIIIIIIIII
1.08	9	IIIIIIIII
1.09	11	IIIIIIIIIII
1.1	17	IIIIIIIIIIIIIIIII
1.11	11	IIIIIIIIIII

1.12	8	IIIIIIII
1.13	14	IIIIIIIIIIIIII
1.14	10	IIIIIIIIII
1.15	3	III
1.16	0	

MEAN = 1.103

STD DEV = 0.0252

Problem 1-2

Specifications:

$LL = 1.06 - .06 = 1.0$

$UL = 1.06 + .06 = 1.12$

Process limits $= \bar{X} \pm 3\sigma = 1.103 \pm 3 \times 0.0252$

Lower limit = 1.0274

Upper limit = 1.1786

$$\frac{1.12 - 1.103}{0.0252} = Z = 0.67$$

From Table A 74.86% is below 1.12 in.

Answer:

25% of production is above the upper spec; 0% is below the lower spec.

Answer:

Shift mean to 1.076 in.
Process limits become: 1.000 in and 1.152 in.

$$\frac{1.12 - 1.076}{0.0252} = Z = 1.75$$

Answer:

From Table A, 96% will be below the upper spec; 4% will be too big.

Chapter Two

Problem 2-1

Answer:

a. $\overline{X} = 32.10$

$\sigma = 0.0185$

b. Process limits $= \overline{X} \pm 3\sigma = 32.10 \pm 3 \times 0.0185$

Lower Limit $= 32.05$

Upper Limit $= 32.16$

c. 0% less than 32.0

d. New mean $= new\, LL + 3\sigma$

New mean $= 32.0 + 3 \times 0.0185$

New mean $= 32.056$

Problem 2-3

Normal dist mean $= 0$, Std dev $= 1$

From Table A, *for* $Z = -2.6$ 0.47%

for $Z = 2.6$ $1 - 99.53\% = 0.47\%$

Problem 2-5

Mean $= 17.10$

Std Dev $= 0.0148$

$\overline{X} + 2.9\sigma = 17.10 + 2.9 \times 0.0148$

Answer:

$$\bar{X} + 2.9\sigma = 17.143$$

$$\bar{X} - 2.9\sigma = 17.10 - 2.9 \times 0.0148$$

$$\bar{X} - 2.9\sigma = 17.057$$

Problem 2-7

Mean = 7850

Std Dev = 880

Answer:

a. $\bar{X} + 3\sigma = 7850 + 3 \times 880$

$$\bar{X} + 3\sigma = 10490$$

$$\bar{X} + 3\sigma = 7850 - 3 \times 880$$

$$\bar{X} + 3\sigma = 5210$$

b. $Z = \dfrac{9000 - 7850}{880} = 1.31$

From Table A, 90.48% of measurements are below 9000, 9.52% are above.

c. $Z = \dfrac{6000 - 7850}{880} = -2.10$

From Table A, 1.79% of measurements are below 6000.

d. $Z = \dfrac{8000 - 7850}{880} = .17$ % below 8000 = 56.75%

$Z = \dfrac{5000 - 7850}{880} = -3.24$ % below 5000 # 0.06%

% between 5000 and 8000 = 56.69%

Problem 2-9

Mean = 1100

Std Dev = 100

a. $\bar{X} + 3\sigma = 1100 + 3 \times 100 = 1400$

$\bar{X} + 3\sigma = 1100 - 3 \times 100 = 800$

b. $Z = \dfrac{1000 - 1100}{100} = -1$

15.87% below 1000 watts, 84.13% above.

c. $Z = \dfrac{950 - 1100}{100} = -1.5$

6.68% below 950 watts.

d. $Z = \dfrac{1200 - 1100}{100} = 1$

84.13% below 1200 watts, 15.87% above.

Chapter Three

Problem 3-1

No	*s*	*a*	*m*	*pl*	*e*	*max*	*min*	*Range*	*Avg*	*std dev*
sample1	48.53	50.04	47.85	48.60	48.70	50.04	47.85	2.19	48.744	0.714
sample2	46.33	48.01	49.81	46.07	48.07	49.81	46.07	3.74	47.658	1.357
sample3	48.74	48.95	47.63	44.93	49.86	49.86	44.93	4.93	48.022	1.701
sample4	47.26	49.39	48.57	47.94	48.15	49.39	47.26	2.13	48.262	0.705
sample5	48.42	48.67	48.31	47.34	50.64	50.64	47.34	3.30	48.676	1.081
sample6	49.19	47.52	49.36	48.73	49.70	49.70	47.52	2.18	48.900	0.757
sample7	47.62	47.72	48.60	49.79	46.25	49.79	46.25	3.54	47.996	1.170

(continued)

Chapter Three Problem 3-1 *(continued)*

No	*s*	*a*	*m*	*pl*	*e*	*max*	*min*	*Range*	*Avg*	*std dev*
sample8	46.95	47.32	47.91	47.65	50.46	50.46	46.95	3.51	48.058	1.243
sample9	47.51	46.94	48.38	48.47	47.36	48.47	46.94	1.53	47.732	0.597
sample10	48.64	47.17	48.81	47.26	47.09	48.81	47.09	1.72	47.794	0.764
sample11	46.07	47.19	47.60	50.66	47.13	50.66	46.07	4.59	47.730	1.550
sample12	48.14	48.79	46.82	46.90	49.57	49.57	46.82	2.75	48.044	1.068
sample13	49.02	47.24	46.31	49.41	48.68	49.41	46.31	3.10	48.132	1.169
sample14	48.18	47.41	48.88	48.16	47.03	48.88	47.03	1.85	47.932	0.648
sample15	47.86	49.20	48.98	47.74	47.48	49.20	47.48	1.72	48.252	0.699
sample16	47.18	48.21	48.79	47.24	47.35	48.79	47.18	1.61	47.754	0.638
sample17	47.35	48.66	48.07	47.62	47.90	48.66	47.35	1.31	47.920	0.444
sample18	46.9	46.52	48.13	46.96	48.60	48.60	46.52	2.08	47.422	0.799
sample19	47.94	48.98	47.12	46.71	49.96	49.96	46.71	3.25	48.142	1.195
sample20	48.1	47.68	47.42	49.15	48.91	49.15	47.42	1.73	48.252	0.676
						TOTAL		52.76	961.422	18.975

Answer:

a. $\overline{\overline{X}} = 48.0711$

b. Sample 5 $\sigma = 1.081$

c. $\overline{R} = 2.638$

$\sigma = 0.94877$

$\sigma_{\overline{X}} = 0.36525$

d. σ' from R $= 1.134$

σ' from $\sigma = 1.129$

e. Process limits from Sig′ from Sig

Upper limit $= 48.07 + 3 \times 1.13 = 51.46$

Lower limit $= 48.07 - 3 \times 1.13 = 44.68$

Problem 3-3

Answer:

a. $Mean = 144.194$

b. $Sample\ 1\ \sigma = 1.941$

c. $\bar{R} = 7.193$

d. $\sigma = 3.092$

e. Upper limit $= 153.47$ Lower limit $= 134.918$

Problem 3-5

$\sigma' = 101.991$

Lower limit $= 2904.03$ Upper limit $= 3515.97$

Problem 3-7

$\sigma' = 0.071$

Lower limit $= 25.9997$ Upper limit $= 26.423$

Approximately 0.16% of containers are below 26 ounces.

Problem 3-9

$\sigma' = 0.005016$

Answer:

Lower limit = 1.0867 Upper limit = 1.1167

Chapter Four

Problem 4-1

X bar Chart *Centerline* = 12.31

UCL = 15.2527

LCL = 9.3673

R Chart *Centerline* = 5.1

UCL = 10.7865

LCL = 0

Problem 4-3

X bar Chart *Centerline* = 12.31

UCL = 15.2307

LCL = 9.3893

Sigma Chart *Centerline* = 1.83

UCL = 3.768

LCL = 0

Problem 4-5

X bar Chart $Centerline = 144.194$

$UCL = 148.344$

$LCL = 140.044$

R Chart $Centerline = 7.193$

$UCL = 15.213$

$LCL = 0$

Problem 4-7

$\bar{X}$ *Chart* $Centerline = 11.25$

$UCL = 11.2983$

$LCL = 11.2017$

R Chart $Centerline = 0.1$

$UCL = 0.2004$

$LCL = 0$

$\sigma'\ est = 0.03946$

79.45% of production is between 11.2 and 11.3 (11.25 $\pm$ 0.05 in.)

Problem 4-9

$\bar{X}$ *Chart* $Centerline = 0.671$

$UCL = 0.67503$

$LCL = 0.66697$

R Chart $Centerline = 0.006978$

$UCL = 0.014754$

$LCL = 0$

Problem 4-11

$\bar{X}$ *Chart* $Centerline = 2.7$

$UCL = 2.716$

$LCL = 2.684$

σ *Chart* $Centerline = 0.01$

$UCL = 0.206$

$LCL = 0$

Lower Process Limit is 2.6643

Upper Process Limit is 2.7357

98.8% of production is between specs.

Chapter Five

Problem 5-1

1/6

Problem 5-3

$$\frac{5}{52}\times\frac{4}{51}\times\frac{3}{50}\times\frac{2}{49}\times\frac{1}{48}\times 48 = 0.0000037$$

Problem 5-5

$$78\times\frac{4}{52}\times\frac{3}{51}\times\frac{48}{50}\times\frac{47}{49}\times\frac{46}{48} = 0.3115$$

Problem 5-7

a. $P(0) = \frac{80}{100}\times\frac{79}{99}\times\frac{78}{98}\times\frac{77}{97}\times\frac{76}{96} = 0.3193$

b. $P(2\ or\ less\ ) = P(0)+P(1)+P(2)$

$$P(1) = 5\times\frac{20}{100}\times\frac{80}{99}\times\frac{79}{98}\times\frac{78}{97}\times\frac{77}{96} = 0.420$$

$$P(2) = 10\times\frac{20}{100}\times\frac{19}{99}\times\frac{80}{98}\times\frac{79}{97}\times\frac{78}{96} = 0.207$$

$P(2\ or\ less\ ) = 0.946$

c. P(3 or more) = 1 − P(2 or less) = 0.054

d. $P(1) = 0.420$

Problem 5-9

a. $P(3w) = 20 \times \frac{10 \times 9 \times 8 \times 80 \times 79 \times 78}{90 \quad 89 \quad 88 \quad 87 \quad 86 \quad 85} = .0158$

b. $P(0w) = 1 \times \frac{80 \times 79 \times 78 \times 77 \times 76 \times 75}{90 \quad 89 \quad 88 \quad 87 \quad 86 \quad 85} = 0.483$

c. $P(0B) = 1 \times \frac{10 \times 9 \times 8 \times 7 \times 6 \times 5}{90 \quad 89 \quad 88 \quad 87 \quad 86 \quad 85} = 0.0000003$

d. $P(3 \text{ or more } B) = 1 - P(2 \text{ or less } B)$

$P(2 \text{ or less } B) = P(0B) + P(1B) + P(2B)$

$P(1B) = 6 \times \frac{80 \times 10 \times 9 \times 8 \times 7 \times 6}{90 \quad 89 \quad 88 \quad 87 \quad 86 \quad 85} = 0.0000324$

$P(2B) = 15 \times \frac{80 \times 79 \times 10 \times 9 \times 8 \times 7}{90 \quad 89 \quad 88 \quad 87 \quad 86 \quad 85} = 0.001066$

$P(3 \text{ or more } B) = 1 - 0.0010985 = 0.9989$

Chapter Six

Problem 6-1

$\bar{X}$ *chart centerline* $= 2$

$UCL = 2.1154$

$LCL = 1.8846$

R *chart centerline* $= 0.2$

$UCL = 0.423$

$LCL = 0$

$\sigma' = 0.085985$

Probability of detecting mean shift to 2.08 is 17.75%

Problem 6-3

$\bar{X}$ *chart centerline* $= 6$

$UCL = 6.2885$

$LCL = 5.715$

R chart centerline $= 0.5$

$UCL = 1.0575$

$LCL = 0$

$\sigma' = 0.21496$

Probability of detecting mean shift to 6.8 is 99.87%

Problem 6-5

$\sigma' = 0.3869$

$n = 9$

Problem 6-7

Answer:

a. *X bar chart centerline* $= 4.5$

$UCL = 4.546$

$LCL = 4.454$

b. *R chart centerline* $= 0.1$

$UCL = 0.169$

$LCL = 0$

$\sigma = 0.03439$

Lower process limit = 4.3968

Upper process limit = 4.6032

New n = 9

Probability of detecting mean shift to 4.55 in a single sample size 9 is 91.18%.

Problem 6-9

$n = 5$

$X\ bar\ chart\ centerline = 6$

$UCL = 6.001342$

$LCL = 5.99866$

$R\ chart\ centerline = 0.00233$

$UCL = 0.00492$

$LCL = 0$

Probability of detecting mean shift to 6.002 in a single sample size 5 is 92.79%.

Chapter Seven

Problem 7-1

$For\ n = 20\ and\ p' = 0.3$

$P(3\ or\ less) = 0.107$

Problem 7-3

$For\ n = 10\ and\ p' = 0.1$

$P(1\ or\ less) = 0.736$

Answer:

P(2 or more) = 0.264

Problem 7-5

For $n = 10$ *and* $p' = 0.08$

$P(0) = 0.4344$

Problem 7-7

$c' = 3$

$P(4 \text{ or less}) = 0.815$

Answer:

P(5 or more) = 0.185

Problem 7-9

$c' = 1$

$P(2 \text{ or less}) = 0.92$

Answer:

P(3 or more) = 0.08

Chapter Eight

Problem 8-1

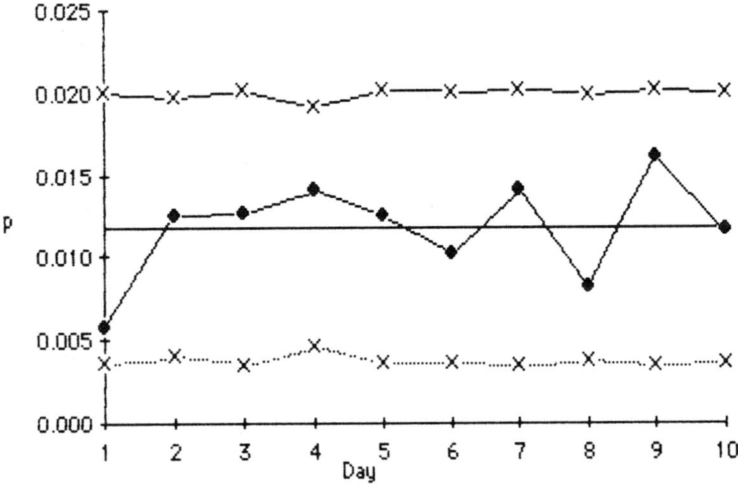

Problem 8.1 p Chart

Problem 8-4

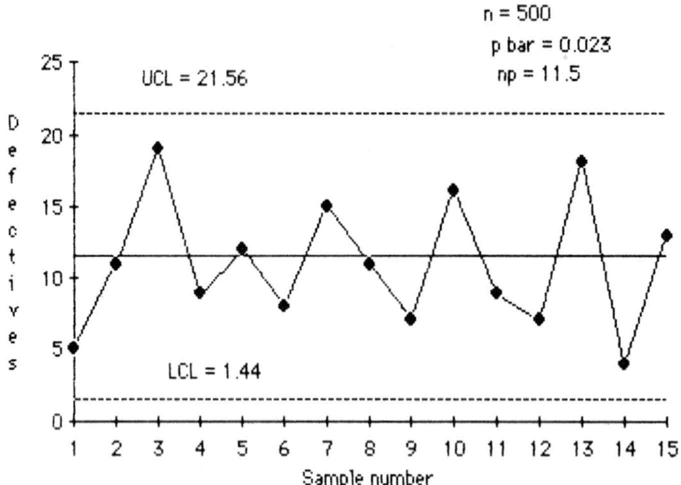

Problem 8.4 np Chart

Problem 8-6

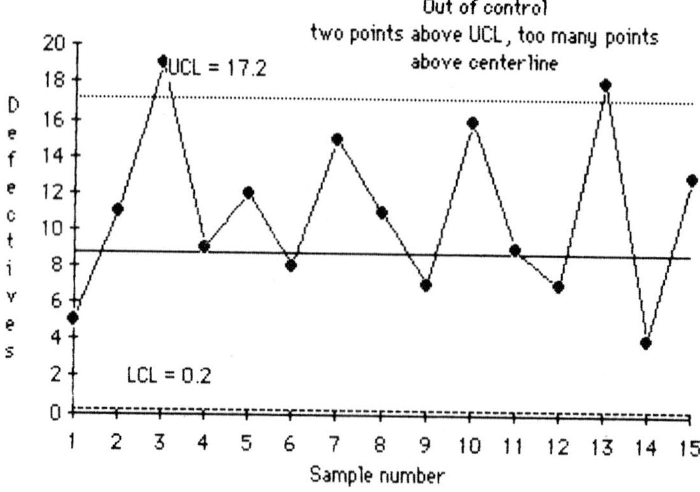

Problem 8.6 nnp Chart

Problem 8-8

$n = 500$

$UCL = 21.56$

With new $p' = 0.05$, $np' = 25$. $P(21\ or\ less) = 0.247$

Answer:

$P(detect) = P(22\ or\ more) = 75.3\%$

Problem 8-10

$n = 100$

$New\ np' = 15$

$UCL = 17.2$

$P(17\ or\ less) = 0.749\ P(detect) = 25.1\%$

Chapter Nine

Problem 9-1

$C = 1, n = 50$

np'	*np'*	*Pa 1st*
50	0.02001	0.736
50	0.04002	0.406
50	0.06003	0.199
50	0.08004	0.092
50	0.10005	0.04
50	0.12006	0.017

Problem 9-3

n	p'	np'	$(c=5)$ Pa	AOQ
200	0.0050	1	0.999	0.005
200	0.0100	2	0.983	0.010
200	0.0150	3	0.916	0.014
200	0.0200	4	0.785	0.016
200	0.0250	5	0.616	0.015
200	0.0300	6	0.446	0.013
200	0.0350	7	0.301	0.011
200	0.0400	8	0.191	0.008
200	0.0450	9	0.116	0.005
200	0.0500	10	0.067	0.003
200	0.0550	11	0.038	0.002
200	0.0600	12	0.02	0.001
200	0.0650	13	0.011	0.001
200	0.0700	14	0.006	0.000
200	0.0750	15	0.003	0.000

Problem 9-4

$n1=100$
$n2=100$
$c1=3$
$c2=4$

n	p'	np'	*Pa 1st*	*P(4or less)*	*P(4 exactly)*	*P(0 on 2nd)*	*Pa 2nd*	*Pa*	*AOQ*
100	0.0100	1	0.981	0.996	0.015	0.368	0.006	0.987	0.010
100	0.0200	2	0.857	0.947	0.09	0.135	0.012	0.869	0.017
100	0.0300	3	0.647	0.815	0.168	0.05	0.008	0.655	0.020
100	0.0400	4	0.433	0.629	0.196	0.018	0.004	0.437	0.017
100	0.0500	5	0.265	0.44	0.175	0.0007	0.000	0.265	0.013

(continued)

n	*p′*	*np′*	*Pa 1st*	*P(4or less)*	*P(4 exactly)*	*P(0 on 2nd)*	*Pa 2nd*	*Pa*	*AOQ*
100	0.0600	6	0.151	0.285	0.134	0.002	0.000	0.151	0.009
100	0.0700	7	0.082	0.173	0.091	0.001	0.000	0.082	0.006
100	0.0800	8	0.042	0.1	0.058	0	0.000	0.042	0.003
100	0.0900	9	0.021	0.055	0.034	0	0.000	0.021	0.002
100	0.1000	10	0.01	0.029	0.019	0	0.000	0.010	0.001

Problem 9-6

$c = 10$

n	*p′*	*np′*	*Pa*	*AOQ*
2000	0.0020	4	0.997	0.002
2000	0.0025	5	0.986	0.0025
2000	0.0030	6	0.957	0.0029
2000	0.0035	7	0.901	0.0032
2000	0.0040	8	0.816	0.0033
2000	0.0045	9	0.706	0.0032
2000	0.0050	10	0.583	0.0029
2000	0.0055	11	0.46	0.0025
2000	0.0060	12	0.347	0.0021
2000	0.0065	13	0.252	0.0016
2000	0.0070	14	0.176	0.0012
2000	0.0075	15	0.118	0.0009
2000	0.0080	16	0.077	0.0006
2000	0.0085	17	0.049	0.0004
2000	0.0090	18	0.03	0.0003

Problem 9-7

$N = 500$
$AQL = 1.5\%$

TABLE	*VALUE*	*SINGLE PLAN*
I	H	
II-A	n = 50	c = 2
V-A	AOQL =	2.43%
X-H	alpha	1.5% = 5% –
VI-A	beta	10%—10%def

Problem 9-9

N = 5000
AQL = 1.5%

TABLE	*VALUE*	*SINGLE PLAN*
I	*L*	
II-A	*n* = 200	*c* = 7
V-A	*AOQL* =	2.1%
X-L	*alpha*	1.5% = 1%
VI-A	*beta*	10%—5.9% *def*

Problem 9-10

Spec: 1.500 *in* ±0.005
N = 150
AQL = 0.40%
TABLE A2 n = *G*
TABLE B-3 n = 15

sample	*meas.*	X^2
1	1.502	2.256004
2	1.497	2.241009
3	1.503	2.259009
4	1.508	2.274064
5	1.501	2.253001
6	1.498	2.244004
7	1.498	2.244004
8	1.503	2.259009
9	1.503	2.259009
10	1.500	2.25

(continued)

sample	*meas.*	X^2
11	1.496	2.238016
12	1.499	2.247001
13	1.501	2.253001
14	1.502	2.256004
15	1.498	2.244004
TOTAL	22.509	33.777139

Line	*info*	*value*
1	n	15
2	Σx	22.509
3	ΣX sqd	33.777139
4	tot x sqd/n	33.7770054
5	ΣX sqd-CF	0.0001336
6	ln5/(n – 1)	9.54E – 06
7	√ln6, s	0.003089
8	X mean	1.5006
9	U	1.505
10	L	1.495
11	(U-mean)/s	1.424
12	(mean-L)/s	1.81
13	Pu table B-5	6.2
14	PL table B-5	6.2
15	Pu + PL	12.4
16	M, table B-3	1.31
17	Reject	12.4 > 1.3

Index